To our wives
Paola, Jane, and Pia
and to our children
Claudio and Benedetta, Julia and Thomas,
Livio and Laura.

In the cover:

Aluminum accumulation in neurons from rats i.p. injected with (2.2 mg Al/Kg) aluminum acetylacetonate (Morin Staining)

P. Zatta and M. Favarato, 1991.

On behalf of the contributors, the Editors would like to thank the Schering-Plough SpA - Italy for the generous support to the organization of *Colloquia Patavina on Aluminum in Chemistry, Biology and Medicine* at the University of Padova and for the publication of this volume which contains the Proceedings of this series of Seminars.

ALUMINUM IN CHEMISTRY BIOLOGY AND MEDICINE

A Series of Advances vol. 1

Edited by

Marino Nicolini · Paolo F. Zatta · Benedetto Corain

cortina international · verona

Raven Press/New York

« Aluminum in Chemistry, Biology and Medicine »
edited by M. Nicolini, P. F. Zatta, B. Corain

Cortina International - Verona
a division of
Edizioni Libreria Cortina - Verona
Via C. Cattaneo 8 - 37121 Verona (Italy)
Tel. 045/594177
Telex 480501 CORTIN I
Telefax 045/597551

ISBN 88-7749-074-8

Contents

Contributors

ALFREY ALLEN C.
Medical Center, 1055 Clermont Street, Denver, CO 80220-USA

BIRCHALL DEREK J.
ICI Research Technology Dept., PO BOX 11, The Heath, Runcorn, Cheshire WA7 4QE-UK

CANDY JOHN M.
MRC Neurochemical Pathology Unit, Newcastle General Hospital, Westgate Road, Newcastle upon Tyne NE4 6BE-UK

COLOMBINI MARCO
Laboratories of Cell Biology, Dept. of Zoology, University of Maryland, College Park, MD 20742-USA

CORAIN BENEDETTO
Dip. di Chimica Inorganica, Metallorganica ed Analitica, Università di Padova, Via Marzolo 1, 35131. Padova-Italy

COSTANTINI SERGIO
Applied Toxicology Laboratory, Istituto Superiore di Sanità, Viale Regina Elena 299, 00161 Roma-Italy

EDWARDSON JAMES A.
MRC Neurochemical Pathology Unit, Newcastle General Hospital, Westgate Road, Newcastle upon Tyne NE4 6BE-UK

FERRIER NICOL I.
MRC Neurochemical Pathology Unit, Newcastle General Hospital, Westgate Road, Newcastle upon Tyne NE4 6BE-UK

GIORDANO ROSA
Applied Toxicology Laboratory, Istituto Superiore di Sanità, Viale Regina Elena 299, 00161 Roma-Italy

HERMAN MARY M.
Depts. of Pathology, Biochemistry and Internal Medicine, University of Virginia, Charlottesville, VA 22903-USA

KATSETOS CHRISTOS D.
Depts. of Pathology and Laboratory Medicine and Neurology, Hahnemann University, Philadelphia, PA-USA

MARTIN BRUCE R.
Chemistry Department, University of Virginia, McCormick Road, Charlottesville, VA 22903-USA

McARTHUR FIONA K.
MRC Neurochemical Pathology Unit, Newcastle General Hospital, Westgate Road, Newcastle upon Tyne NE4 6BE-UK

McKEITH IAN G.
MRC Neurochemical Pathology Unit, Newcastle General Hospital, Westgate Road, Newcastle upon Tyne NE4 6BE-UK

McLAUGHLIN IAN
MRC Neurochemical Pathology Unit, Newcastle General Hospital, Westgate Road, Newcastle upon Tyne NE4 6BE-UK

MORRIS CHRISTOPHER M.
MRC Neurochemical Pathology Unit, Newcastle General Hospital, Westgate Road, Newcastle upon Tyne NE4 6BE-UK

MOUNTFORT SIMON A.
MRC Neurochemical Pathology Unit, Newcastle General Hospital, Westgate Road, Newcastle upon Tyne NE4 6BE-UK

NICOLINI MARINO
Dip. di Scienze Farmaceutiche, Università di Padova, Via Marzolo 5, 35131 Padova-Italy

OAKLEY ARTHUR E.
MRC Neurochemical Pathology Unit, Newcastle General Hospital, Westgate Road, Newcastle upon Tyne NE4 6BE-UK

SAVORY JOHN
Depts. of Pathology, Biochemistry and Internal Medicine, University of Virginia, Charlottesville, VA 22903-USA

TAYLOR GEOFFREY A.
MRC Neurochemical Pathology Unit, Newcastle General Hospital, Westgate Road, Newcastle upon Tyne NE4 6BE-UK

WARD M.K.
MRC Neurochemical Pathology Unit, Newcastle General Hospital, Westgate Road, Newcastle upon Tyne NE4 6BE-UK

WILLS MICHAEL R.
Depts. of Pathology, Biochemistry and Internal Medicine, University of Virginia, Charlottesville, VA 22903-USA

WISNIEWSKI HENRYK M.
New York State Institute for Basic Research in Developmental Disabilities, 1050 Forest Hill Road, Staten Island, NY 10314-USA

ZATTA PAOLO F.
CNR-Centro per lo Studio della Biochimica e della Fisiologia delle Emocianine ed altre Metallo-Proteine, Dip. di Biologia, Università di Padova, Via Trieste 75, 35131 Padova-Italy

Preface

> It is not his possession of knowledge,
> or irrefutable truth, that makes the man of
> science, but his persistent and critical quest
> for truth.
>
> *Karl Popper*

Recently five books dealing with the relevance of aluminum to biology and human health were offered to the attention of the scientific community (1-5).

Why one more book? The rate of production of scientific papers on aluminum biology in the last two decades is so large that a continuous updating in this field is highly advisable. To this end, we have organized a permanent forum, i.e. "Colloquia Patavina on Aluminum in Chemistry, Biology and Medicine", with the aim to let international prominent scholars active in this field to present their recent research work in a critical and organized form.

The contributors of this book were invited to give lectures at the University of Padova, Italy from March to May 1991, and to kindly materialize their contributions in the form of the chapters presented herein.

We strongly hope to repeat this cultural experience in 1993, probably as a two-three day workshop, and to publish a second volume of selected lectures.

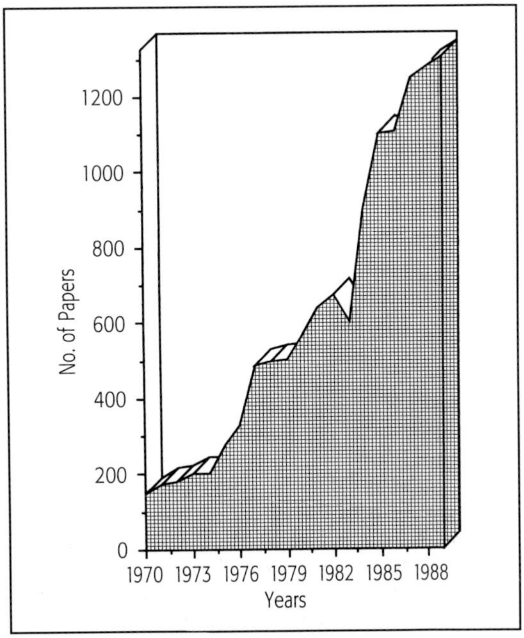

The meeting in Padova was scientifically very successful and stimulating, and gave us the opportunity to build up the following proposal aimed at contributing to the solution of a major problem of aluminum biology.

In fact, we all are aware that the most exciting and undoubtedly interesting issue in aluminum biology is the relevance of abnormal metal accumulation in selected human brain areas in Alzheimer's disease. We are also aware that controver-

sial findings on this fundamental phenomenological datum are widely described in the relevant literature. We think it is time to activate an "Interlaboratory Project" aimed at closing the controversy on the aluminum presence and compartmentation in Alzheimer's disease, and on the characteristic features such as neurofibrillary tangles and senile plaques.

The involved laboratories should be requested to exchange biological material and standards suitable for more analytical methods. We are looking forward to receiving proposals on this matter and preparing a list of candidates interested in this challenge. The final evaluation and reliability of the results should be left to a panel of prominent experts belonging to outstanding international institutions such as N.I.H., I.U.P.A.C., Chemical Societies etc.

The unambiguous, internationally controlled resolution of this controversy will represent a milestone in the understanding of the molecular bases of this challenging disease.

The editors wish to thank all the authors of contributions for their remarkable commitment in writing their papers and their readiness.

They also thank the Publisher, Cortina International of Verona, and the Printer, Bertoncello Artigrafiche, for their considerable efforts that have led to this excellent editorial achievement.

Padova September 1991

Marino Nicolini
Paolo F. Zatta
Benedetto Corain

1. Aluminum and its Role in Biology, H. Sigel and A. Sigel, Eds., Marcel Dekker, Inc. New York 1988.
2. Aluminum in Food and the Environment, R.C. Massey and D. Taylor, Eds., The Royal Society of Chemistry, Whitstable-Kent, 1989.
3. The Environmental Chemistry of Aluminum, G. Sposito, Ed., CRC Boca Raton, 1989.
4. Environmetal Chemistry and Toxicology of Aluminum, T.E. Lewis, Ed., Lewis Publishers, Inc., Chelsea-Michigan, 1989.
5. Aluminum and Health, H.J. Gitelman, Ed., Marcel Dekker Inc., New York, 1989.

Section I

CHEMISTRY

Aluminum in Biological Systems

R.B. Martin

1. Introduction

Comprising 8% of the earth's crust, aluminum is locked in minerals that conceal its status as the most abundant metal and the third most abundant element after oxygen and silicon. It occurs only in combined form: as an oxide in bauxite, the primary ore, and in complex aluminosilicates such as micas and feldspars. In contrast to its abundance in the crust, the ocean concentration is below 1 μg Al per liter; this low level may be the result of the accumulation of aluminum and silicon by diatoms.[1-3] Most natural waters contain insignificant amounts of aluminum, except for some volcanic regions and alum springs. Any freed Al^{3+} is usually deposited in sediment as a hydroxide. With the advent of acid rain, metal ions such as aluminum, mercury, and lead escape from mineral deposits and appear in fresh waters. Acid rain serves as the key that springs the lock for metal ion release. The Al^{3+} concentration increases sharply in clear water lakes at $pH < 6$ where micromolar amounts may occur.[4] It is more damaging to fish than increased acidity; they can tolerate only nanomolar Al^{3+} levels.

Plants, such as tea, that accumulate Al(III) do so in acidic soils and evidently detoxify the Al(III) by storing a chelated version in cell vacuoles of older leaves separate from the more metabolically active parts of the plant. Tea plants have been found with as much as 3% Al(III) in older leaves and only 0.01% in younger ones, a 300-fold difference.[5] Typical tea infusions contain about 50 times as much Al(III) as do infusion from coffee.[6] Addition of milk to tea should immobilize Al(III) as an insoluble phosphate, while addition of lemon will strongly complex the Al(III) in a dangerous citrate complex (Sections 5 and 7).[7-10]

Grass tetany is an acute muscle spasmodic condition of ruminant animals (usually females stressed by lactation) grazing on lush green pastures in the early spring and resulting in convulsive death unless treated. It is accompanied by low Mg^{2+} in growing grass and blood serum, but this deficiency does not suffice to produce the condition. In 1971 it was suggested to be due to high Al(III) and low Mg^{2+} in grass forage, and a phosphate fertilizer with a Mg^{2+} supplement was recommended as a cure.[11] Like some other diseases with a claimed linkage to Al(III), this one has proved controversial and the question unresolved.[12]

Coupled with the increasing availability of Al^{3+} is its identification as a causative or associative agent in several kinds of human disorders.[13,14] Al(III) is the acknowledged causative agent in conditions arising from dialysis: vitamin D-resistant osteomalacia, iron adequate microcytic anemia, and dialysis dementia. A recent study of 10^4 long-term hemodialysis patients finds that a serum Al(III) concentration of greater than 1.5 μM increases the mortality risk, and 40% of the patients exceed this level.[15] In normal individuals the serum Al(III) level is less than 0.4 μM. Upon acute ingestion even normal adults may accumulate Al(III) in bones.[16] Whether Al(III) causes or associates with other neurodegenerative diseases, such as Alzheimer's disease, is a matter discussed in other chapters of this book and in review articles and books mentioned in the references 9, 10, 17, 18.

The chemistry of Al^{3+} is relatively simple. It reacts 10^7 times faster than Cr^{3+}, is much more soluble than Fe^{3+}, and exhibits only one oxidation state in biological systems, Al^{3+}. Metallic Al is too reactive to be found free in nature, and the metal is won from its ores only with difficulty.[9] Thus there is no oxidation-reduction chemistry to Al^{3+} in biology. The chemistry of aqueous Al^{3+} is made difficult, however, by hydrolytic equilibria in aqueous solutions, to which subject we turn in the next section.

The effective ionic radius of Al^{3+} in six-fold coordination is 54 pm. By way of comparison other values are Ga^{3+}, 62; Fe^{3+}, 65; Mg^{2+}, 72; Zn^{2+}, 74; Fe^{2+}, 78; and Ca^{2+}, 100 pm.[19] On the basis of the radii, though quite small, Al^{3+} is closest in size to Fe^{3+} and Mg^{2+}. Ca^{2+} is much larger, and in its favoured eight-fold coordination exhibits a radius of 112 pm, yielding a volume nine times greater than Al^{3+}. For this and other reasons we have argued that in biological systems Al^{3+} will be competitive with Mg^{2+}, rather than Ca^{2+}.[9,20] Both Al^{3+} and Mg^{2+} favor oxygen donor ligands, especially phosphate groups.[21] *Wherever there is a process involving Mg^{2+}, seek there an opportunity for interference by Al^{3+}.*

2. Hydrolytic Equilibria

Whatever ligands may be present, understanding the state of Al(III) in any aqueous system demands awareness of the species that Al(III) forms with the components of water at different pH values. In solutions more acid than pH < 5, Al(III) exists as an octahedral hexahydrate, $Al(H_2O)_6^{3+}$, usually abbreviated as Al^{3+}. As a solution becomes less acidic, $Al(H_2O)_6^{3+}$, undergoes successive deprotonations to yield $Al(OH)^{2+}$, $Al(OH)_2^+$, and soluble $Al(OH)_3$, with a decreasing and variable number of water molecules.[9,22] Neutral solutions give an $Al(OH)_3$ precipitate that redissolves, owing to formation of tetrahedral aluminate, $Al(OH)_4^-$, the primary soluble Al(III) species at pH > 6.2. (Since we are concerned with relatively low concentrations of total Al(III) in biological systems in the presence of other ligands, we do not consider formation of polynuclear species, compositions of which are time dependent[23]).

Equilibria among the free metal ion and four hydrolyzed species are described in four reactions. As usual we abbreviate the hexahydrate occuring in acid solutions as Al^{3+}. This abbreviation disguises the actual reaction: deprotonation of a water bound metal ion according to

$$Al(H_2O)_6^{3+} + H_2O \rightleftarrows Al(H_2O)_5(OH)^{2+} + H_3O^+$$

By common consent we rewrite this reaction as a hydrolysis.

$$Al^{3+} + H_2O \rightleftarrows Al(OH)^{2+} + H^+$$
$$K_1 = (H^+)[Al(OH)^{2+}]/[Al^{3+}] \qquad (1)$$

This reaction sets off successive deprotonations that, when written as hydrolysis reactions, yield the following reactions and equilibrium constants.

$$Al(OH)^{2+} + H_2O \rightleftarrows Al(OH)_2^+ + H^+$$
$$K_2 = (H^+)[Al(OH)_2^+]/[Al(OH)^{2+}] \qquad (2)$$

$$Al(OH)_2^+ + H_2O \rightleftarrows Al(OH)_3 + H^+$$
$$K_3 = (H^+)[Al(OH)_3]/[Al(OH)_2^+] \qquad (3)$$

$$Al(OH)_3 + H_2O \rightleftarrows Al(OH)_4^- + H^+$$
$$K_4 = (H^+)[Al(OH)_4^-]/[AlOH_3] \qquad (4)$$

In the equilibrium constant expressions, the brackets represent concentrations, and the parentheses, activity on the usual pH scale standardized with NBS buffers.[24] Because of the insolubility of Al(III) hydroxides, determination of the constants K_1 to K_4 has proven difficult. We take $pK_1=5.5$, $pK_2=5.8$, $pK_3=6.0$, and $pK_4=6.2$. These values come from a tabulation at 0.6 ionic strength on a concentration scale for H^+;[25] conversion to 0.16 ionic strength and the usual activity pH scale is partially offsetting. (The unusual closeness of the four pK_a values has been ascribed to a decrease in coordination number,[9] in strong contrast with Fe^{3+} [22]).

The upper half of figure 1 shows the distribution of free metal ion and mononuclear hydrolyzed species based on the four successive pK_a values given above. The mole fraction distribution curves in the upper part of figure 1 combine the four acidity constant expressions 1 – 4 with the mass balance equation for total soluble metal ion in all its forms.

$$T_{Al} = [Al^{3+}] + [Al(OH)^{2+}] + [Al(OH)_2^+] + [Al(OH)_3] + [Al(OH)_4^-]$$

Since only mononuclear terms appear, the distribution curves are independent of the total concentration. The octahedral hexahydrate $Al(H_2O)_6^{3+}$ dominates at pH < 5, and the tetra-

Fig. 1. Al^{3+} hydrolysis. Upper half: mole fraction of soluble species as a function of pH. Lower half: negative logarithm of molar concentration in saturated solutions for the free ion, $[Al^{3+}]$, straight dashed line; and the sum over all species present, T_{Al} curved line. The straight dotted line represents the concentration of tetrahedral aluminate ion, $[Al(OH)_4^-]$.

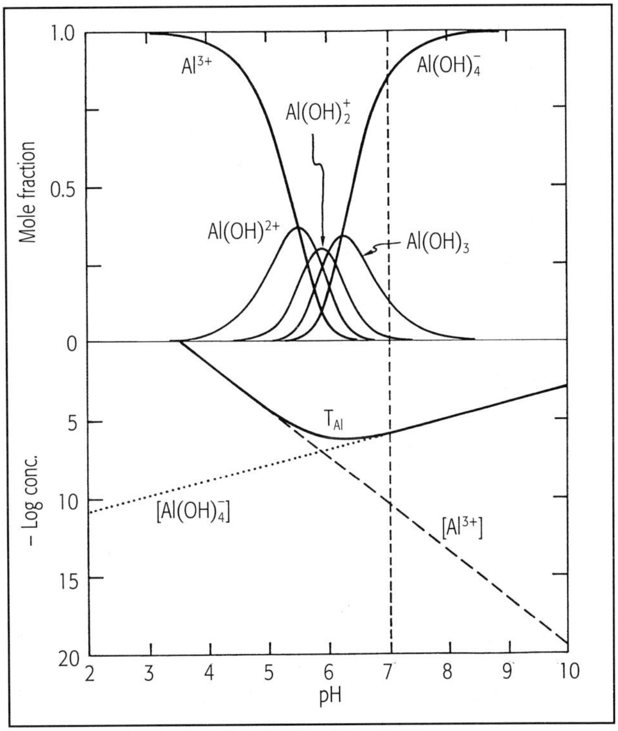

hedral $Al(OH)_4^-$ at pH >7, while there is a mixture of hydrolyzed species and coordination numbers between $5 < pH < 7$.[9,22]

To this point we have described the equilibria and resulting distribution among soluble, mononuclear Al(III) species without considering the absolute amounts permitted by the limited solubility of $Al(OH)_3$. The solubility of $Al(OH)_3$ may be described as

$$Al(OH)_3(s) + 3H^+ \rightleftarrows Al^{3+} + 3H_2O$$
$$K_s = [Al^{3+}]/(H^+)^3 \qquad (5)$$

At reasonable temperatures the stable crystalline phase is Gibbsite, a-$Al(OH)_3$, with log $K_s = 9.2$ under our conditions.[9,26] In aqueous solutions, however, Gibbsite reaches equilibrium slowly, sometimes only after years, and in biological systems equilibrium with respect to Gibbsite rarely occurs. The presence of organic ligands favors formation of non-crystalline $Al(OH)_3$. Therefore, we employ the thirty times greater K_s value (log $K_s = 10.7$), appropriate for amorphous precipitates that form relatively rapidly.[23,26]

The lower half of figure 1 applies to solutions saturated with amorphous $Al(OH)_3$. The dashed straight line of slope 3 in the lower half of the figure gives the molar concentration of the free metal ion, $[Al^{3+}]$. The solid curve represents the total concentration of the free metal ion and all mononuclear hydrolyzed forms, T_{Al}, with the distribution depicted in the upper half of the figure. From the lower part of the figure we learn that large amounts of $Al(OH)_3$ dissolve in acidic stomachs.

Figure 1 also shows as a dotted line the molar concentration of aluminate, $[Al(OH)_4^-]$, the main species at pH > 6.2. At any pH the ordinate distance between the straight lines for $[Al(OH)_4^-]$ and $[Al^{3+}]$ gives the logarithm of their concentration ratio. Thus at pH 7.0 the molar ratio of $[Al(OH)_4^-]/[Al^{3+}]$ is given by $10^{4.5} = 3 \times 10^4$. Therefore, $Al(OH)_4^-$ (aluminate) should be the starting point for thinking about aluminum in biological systems.

As with any other ligand, hydroxide by reactions 1 – 5 withdraws Al^{3+} from solution. In aqueous solutions, regardless of the other ligands present, reactions 1 — 4 and the species distribution of the upper half of figure 1 apply. These equilibria must be considered in all solutions containing Al(III). If in addition other ligands are incapable of holding Al(III) in solution, it becomes necessary to include the solubility equilibrium of equation 5.

What happens when an $AlCl_3$ solution is administered at a local concentration of 0.01 M to a tissue at pH 7? Ascension of the dashed pH 7 line in figure 1 indicates that the permissible free Al^{3+} is only $10^{-10.3}$ M and that of all soluble forms is $10^{-5.7}$ M=2 μM. Unless the remainder of the added Al(III) has been complexed by other ligands, it will form insoluble $Al(OH)_3$. From the upper part we see that of the soluble forms 85% is $Al(OH)_4^-$ and 14% $Al(OH)_3$. Of the Al(III) administered at pH 7 only 2 μM appears in soluble forms, most of which is $Al(OH)_4^-$ since most of the added Al(III) precipitates or coordinates to nearby ligands. To keep Al(III) in solution it may be administered as a complex.

3. Phosphate Binding

It is to phosphate groups that Al^{3+} will be frequently and significantly complexed in biological systems. We now compare the binding strengths of various phosphate containing ligands. Al^{3+} binds 560 times more strongly than Ca^{2+} to the surface of phosphatidylcholine vesicles with a stability constant log K_s= 3.2.[27] Stability constant logarithms recently determined for Al^{3+} binding to adenosine-5′-nucleotides[28] and 2,3-diphosphoglycerate (DPG)[29] appear in Table I. The strongest stability constants appear for ADP and ATP where chelation occurs. At about 3 mM there is three times as much DPG as ATP in the blood red cell but, since ATP binds Al^{3+} 80 times strongly, most Al^{3+} binds to ATP rather than to DPG. To within one standard deviation, equal concentrations of Al^{3+} appear on both sides of the erythrocyte membrane.[30] For comparison, the stability constant for Mg^{2+} binding to ATP and other nucleoside triphosphates is log K_1= 4.3,[31] 4000 times weaker than for Al^{3+}. Thus, 0.2 μM Al^{3+} competes with 1 mM Mg^{2+} for ATP.

We use the results for AMP and DPG to evaluate the stability constants for Al^{3+} binding to simple phosphates, R-OPO_3^{2-}. For such ligands and several metal ions (other than Al^{3+}), log K_s versus pK_a plots are linear with a low slope of about 0.3.[32] It has been argued that in DPG the pK_a=6.1 should be assigned mainly to the 3-phosphate and the pK_a=7.2 to the 2-phosphate.[29] Owing to the low slope, in the $HLAl^+$ species, most complexes bear the proton at the more basic 2-phosphate and the Al^{3+} at the 3-phosphate. With phosphate pK_a=6.1 both AMP and DPG also exhibit similar stability constants of log K_1=6.2 and 6.0 and identical log K_2=4.2 (Table I). These concordant values serve as our basis for estimating stabili-

Table I. **Al^{3+} binding to phosphate ligands[a].**

Ligand	pK_a^b	Log K_1	Log K_2
ATP	6.4	7.9	4.6
ADP	6.3	7.8	4.3
AMP	6.1	6.2	4.2
2,3-DPG	6.1,7.2	6.0($AlLH^+$)	4.2 ($AlL_2H_2^-$)[c]
$HOPO_3^{2-}$	6.77	6.3[d]	4.4[d]

a. Adenosine phosphates from ref. 28 and 2,3-diphosphoglycerate from ref. 29.
b. Phosphate group values adjusted to usual activity pH scale.
c. 2,3-DPG 2:1 complex proton losses with pK_a=5.7 and 6.4 so that main species in neutral solutions is AlL_2^{3-}.
d. Estimated, see text.

ty constants to other R-OPO_3^{2-}. Hydrogen phosphate, $HOPO_3^{2-}$, accepts a proton with $pK_2 = 6.77$, and based on a slope of 0.3, we estimate for $Al^{3+} + HOPO_3^{2-} \rightarrow HOPO_3Al^+$, log $K_1 = 6.3$, and for a second ligand, log $K_2 = 4.4$. The former value agrees with one experimental value of 6.3[33] and is significantly less than a second of 7.7.[34] Since experimental agreement is poor and the experiments difficult owing to the insolubility of $AlPO_4$, we utilize the 6.3 value. (The conditional stability constants and pAl values appear in the seventh row of Table II below).

There is a widespread supposition that Al^{3+} binds to DNA in the cell nucleus. This conclusion derives from several *in vivo* and *in vitro* studies. Aluminum has been found in the nuclear region of brain neuronal cells that contain neurofibrillary tangles, both in Alzheimer's disease victims[35] and in the indigenous population of Guam that suffers from amyotrophic lateral sclerosis and parkinsonism-dementia.[36] More detailed studies located the aluminum in the chromatin of brain neuronal nuclei, both in Alzheimer's victims[37] and in animals subjected to CNS injection of aluminum.[37-39] An *in vitro* study of Al^{3+} interactions with DNA employed several physical methods from pH 5.0 to 7.5.[40] Severe competition from hydroxo complex formation indicates only weak Al^{3+} association with DNA in neutral solutions. Moreover, this study was conducted at high Al(III) to DNA phosphorus mole ratios > 0.3 (more than ten times that found in Alzheimer's brains[37]) and in the presence of only a 5 mM $NaNO_3$ background. Lower Al(III) and higher ionic strengths maintained with some dipositive ions such as Mg^{2+} would greatly weaken the Al(III) interactions. Polynucleotide cross-linking by Al(III) occurs only at pH < 6 with high mole ratios > 0.4, and inhibits renaturation of denatured DNA.[40-42]

Table II. **Negative logarithm of free Al^{3+} concentration, pAl[a].**

Ligand or Inorganic System[b]	**Log K_s**	**pH 6.6**		**pH 7.4**	
		Log $K_{6.6}$	**pAl**	**Log $K_{7.4}$**	**pAl**
DNA	< 5.6	< 5.6	< 7.3	< 5.6	< 7.3
F^-, 3 mM, with OH^-	6.4,5.2		8.6		8.9
Salicylate, 0.2 mM	12.9,10.6	6.3,4.0	9.1	7.1,4.8	10.7
Amorphous $Al(OH)_3$	insoluble		9.1		11.5
Al(III), $pK_a = 5.5, 5.8, 6.0, 6.2$			9.1		12.1
Catecholamines	15.6,13.0	7.4,4.8	9.7	9.0,6.4	12.8
$HOPO_3^{2-}$	6.3	5.9	9.9[c]	6.2	9.5[d]
Kaolinite[e]	insoluble		10.2		12.6
$AlPO_4$	insoluble		11.7[c]		12.1[d]
Nitrilotriacetate (NTA)	11.1	10.0	11.7	11.6	13.3
2,3-DPG, 3mM	12.5	11.6	12.2	12.2	13.1
ATP, 1mM	7.9,4.6	8.9	12.3	9.8	13.0
Citrate, 0.1 mM	8.1	11.3	13.3	12.7	14.7
Transferrin				13.6,12.8	15.3
EDTA	16.2	13.1	14.8	14.7	16.4
Deferoxamine	24.1	16.8	18.4	19.2	20.8

a. 1 μM total Al(III) except for insoluble salts. Equilibria in addition to those related to listed log K_c values often required to calculate pAl.
b. 50 μM ligand unless otherwise noted.
c. 10 mM total phosphate.
d. 2 mM total phosphate.
e. $Al_2(OH)_4Si_2O_5$ with 5 μM $Si(OH)_4$, typical of plasma.

Binding of Al^{3+} to DNA may be placed on a more quantitative basis. Potentiometric titration with a fluoride selective indicator electrode and equilibrium dialysis have both been used to assess quantitatively the tendency for Al^{3+} binding to DNA at pH 4.5, 5.0, and 5.5 in the presence of 1 to 100 mM NaCl.[43] Higher pH values were not possible because DNA binding does not compete with metal ion hydrolysis and precipitation (Table II below). The interpretation assumes (but not critically) that each bound Al^{3+} displaces three Na^+. (The derived results show that the stability constant for Al^{3+} binding decreases with increasing ionic strength in a manner more consistent with the displacement of one rather than three Na^+). At the highest ionic strength of 100 mM NaCl, the stability constant for Al^{3+} binding to DNA is log $K_s = 5.6$; this value becomes the upper limit under intracellular conditions. Since there is no accompanying protonic equilibria, this value also represents the maximum conditional stability contant in neutral solutions. The 5.6 value is more than 3.3 log units less than the conditional stability constant for ATP at pH 6.6 (Table II). Therefore, DNA cannot compete with ATP, $HOPO_3^{2-}$, and other ligands in Table II for Al^{3+}. With very weakly basic phosphates of $pK_a \sim 1$, DNA serves merely as a polyelectrolyte interacting with Al^{3+} weakly and non-specifically. Residing at the very top of Table II with the lowest pAl values or highest allowed free Al^{3+} concentrations DNA loses Al^{3+} to all other entries in the Table. We find that Al^{3+} binding to DNA is so weak under intracellular conditions that it is not competitive by several orders of magnitude with either metal ion hydrolysis or insolubility of even an amorphous $Al(OH)_3$. No matter what other ligands are present, this competition with aqueous solvent components will remain. *Therefore, we conclude that the observation of aluminum with nuclear chromatin is not due to its coordination to DNA but to other ligands.*

What ligands might be binding Al^{3+} in the cell and especially in the nuclear chromatin region? ATP and ADP are comparably strong Al^{3+} binders. A crucial Al^{3+} binding site in chromatin promises to be phosphorylated proteins, perhaps phosphorylated histones. Phosphorylation and dephosphorylation reactions normally accompany cellular processes. The phosphate groups of any phosphorylated protein provide the requisite basicity, and in conjuction with juxtaposed carboxylate or other phosphate groups, become strong Al^{3+} binding sites. Abnormally phosphorylated proteins have been found in Alzheimer's disease brains.[44, 45] Al^{3+} aggregates highly phosphorylated brain cytoskeletal proteins.[46] Ternary Al^{3+} complexes have received little study,[47] and Al^{3+} has been used as a tanning or crosslinking reagent. There is a strong possibility that Al^{3+} crosslinks proteins, and proteins and nucleic acids in the nucleus.

In the human body extracellular fluids contain about 2 mM total phosphate at pH 7.4 and intracellular fluids about 10 mM total phosphate at pH 6.6. Al(III) forms an insoluble salt with phosphate often designated as $AlPO_4$ or sometimes as $AlPO_4 \cdot 2H_2O$, corresponding to the composition of the mineral variscite. The solubility product constant has recently been determined at 37° C and 0.15 M ionic strength and reported[34] as $[Al^{3+}][PO_4^{3-}] = K_{sp} = 10^{-19.1}$. At 0.16 M ionic strength the usual mixed pK_a values for successive deprotonations from $H_3PO_4 \rightarrow H_2PO_4^- \rightarrow HPO_4^{2-} \rightarrow PO_4^{3-}$ are $pK_1 = 2.0$, $pK_2 = 6.77$, and $pK_3 = 11.6$.[48] There is little temperature dependence to any of the three successive ionizations from phosphoric acid. The concentration of the most basic form is given by

$$[PO_4^{3-}] = T_P/(\{[(H^+)/K_1+1](H^+)/K_2+1\}(H^+)/K_3+1)$$

where T_P is the total inorganic phosphate concentration in all four forms and the pK_a values are given above. Combining the last two equations we obtain for the $3 < pH < 11$ range, where there are only two mains forms, $H_2PO_4^-$ and HPO_4^{2-}, the reduced expression

$$[Al^{3+}] = K_{sp}(H^+)[(H^+)+K_2]/K_2K_3T_P = \\ =(H^+)[(H^+)+10^{-6.77}]/5T_P \qquad (6)$$

Equation 6 furnishes an estimate of the free Al^{3+} concentration at 37 °C allowed by the solubility of variscite at 0.16 M ionic strength for a designated $3 < pH < 11$ and total phosphate concentration T_P. For intracellular fluids at pH 6.6 containing 10 mM total phosphate, equa-

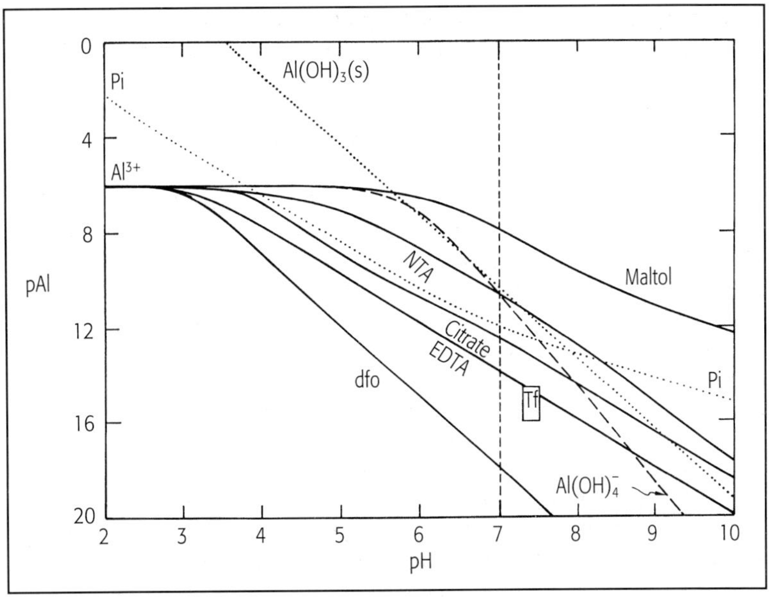

Fig. 2. Negative logarithm of free aqueous Al^{3+} concentration versus pH allowed by several Al(III) complexes and solids. Dotted straight line of slope 3 represents amorphous $Al(OH)_3(s)$ and dotted curve 4.5 mM inorganic phosphate. Dashed and all soid curves refer to 1 μM total Al(III) beginning as Al^{3+} at pAl= 6 at low pH on the left hand ordinate. Dashed curve shows decrease of free Al^{3+} due to metal ion hydrolysis leading to $Al(OH)_4^-$ at bottom right. Solid curves (with mole ratios) represent maltol (3 : 1), NTA (2 : 1), citrate (4 : 1), EDTA (2 : 1), and deferoxamine, dfo (2 : 1). The rectangle labeled Tf refers to 50 μM transferrin under blood plasma conditions.

tion 6 yields $-\log [Al^{3+}] = pAl = 11.7$, while for extracellular fluids at pH 7.4 containing 2 mM total phosphate pAl=12.1. This pair of values appears in the ninth row of Table II below and represents extremely low maximum free Al^{3+} concentrations. Under the same pair of conditions the free Ca^{2+} concentrations permitted by insoluble hydroxyapatite, $Ca_5(PO_4)_3OH$, a principal constituent of bones and teeth, are nearly 10^7 times greater.[48] Thus thermodynamically, less than nanomolar levels of Al^{3+} displace Ca^{2+} from phosphate binding. Even at low accumulation levels Al(III) impairs bone mineralization.[49]

The curve labelled P_i in figure 2 below describes pAl as a function of pH according to equation 6 for an inorganic phosphate concentration of 4.5 mM, the geometric mean of and extra- and intracellular concentrations. The curve therefore lies within 0.35 log units of the 2 and 10 mM concentrations for extra- and intracellular phosphate.

4. Exchange

In addition to stability of metal ion complexes an important and often overlooked feature is the rate of ligand exchange out of and into the metal ion coordination sphere. Ligand exchange rates take on a special importance for Al^{3+} because they are slow, and systems may not be at equilibrium. The rate for exchange of inner sphere water with solvent water has been measured for many metal ions, and the order of increasing rate constants in acidic solutions is given by

$$Al^{3+} \ll Fe^{3+} < Ga^{3+},\ Be^{2+} \ll Mg^{2+} < Fe^{2+} < \\ < Zn^{2+} < Ca^{2+}$$

Each inequality sign indicates an approximate 10-fold increase in rate constant from 1.3 sec^{-1} for Al^{3+} and increasing through 8 powers of 10 to 10^8 sec^{-1} for Ca^{2+} at 25 °C.[50, 51] Though these specific rate constants refer to water exchange in aquo metal ions, they also reflect relative rates of exchange of other unidentate ligands. Reducing Fe^{3+} to Fe^{2+} gains a 10^4 fold rate increase. Chelated ligands exchange more slowly, but the order is maintained. The slow ligand exchange rate for Al^{3+} renders it useless as a metal ion engaged in enzyme active site reactions. The 10^5 times faster rate for Mg^{2+} furnishes sufficient reason for Al^{3+} inhibition of enzymes with Mg^{2+} cofactors. Processes involving rapid Ca^{2+} exchange would obviously be totally thwarted by substitution of the 10^8 fold slower Al^{3+}.

A comparison of the aqueous coordination chemistries of Al^{3+} and Fe^{3+} emphasizes rate effects.[52]

As mentioned, the above rate constants refer to aquo metal ions in acidic solutions; for example, to hexacoordinate $Al(H_2O)_6^{3+}$. Partially hydrolyzed Al(III) often undergoes faster substitution; for example, salicylates[53, 54] and deferoxamine[55] react about 10^4 times faster with $Al(OH)^{2+}$ than with $Al(H_2O)_6^{3+}$. Such rate enhancements have been attributed to loosening of the metal ion coordination sphere or to the metal ion bound OH^- serving as an internal base in deprotonating the incoming ligand, both in terms of a dissociative pathway of a hexacoordinate metal ion. Based on the equilibrium description in Section 2, we suggest that the rate acceleration results from addition of an associative pathway that becomes possible with the reduced coordination numbers of hydrolyzed Al(III) . Unfortunately, no rate studies appear to have been performed on tetrahedral $Al(OH)_4^-$, the main aquo species in neutral solutions. The coordination number of Al(III) complexes in many biological systems remains uncertain, and the problem is compounded when more than one kind of ligand binds to Al(III). When a reduced coordination number does occur, the relative rate for ligand exchange on Al(III) may increase, but it is likely still at least 100 fold slower than for Mg^{2+}. (Hg^{2+} and CH_3Hg^+ make linear complexes with sulfhydryl groups so strong (mercaptan = mercury capture) that the observed rapid metal ion exchange among thiol ligands cannot be accomodated by a dissociative pathway. Required is an associative pathway with an intermediate of higher coordination number[19]).

5. Ligands for Al^{3+}

The next three sections being with a qualitative overview of Al^{3+} binding to a variety of molecules, continue in Section 6 with a quantitative summary, and conclude in Section 7 with some details of the calculations.

The most likely Al^{3+} binding sites are at oxygen atoms, especially if they are negatively charged. Carboxylate, catecholate, and phosphate groups are the strongest Al^{3+} binders. Amines are not strong Al^{3+} binders except as part of multidentate ligand systems such as nitrilotriacetate (NTA) and EDTA. A variety of multidentate chelates have been compared for their ability to bind Al^{3+} and Ga^{3+}.[56] The nitrogenous bases of DNA and RNA do not bind Al^{3+}.[7, 9, 10, 57] Sulfhydryl groups, as demonstrated by the sulfhydryl laden protein metallothionein,[58] do not bind Al^{3+}.

5.1. Simple Carboxylates

Al^{3+} forms only weak complexes with simple carboxylate ligands. Careful potentiometric studies establish that even excess acetic and propionic acids do not yield mononuclear complexes but only a single binuclear species, $(RCOO^-)Al_2(\mu\text{-}OH)_2^{3+}$, with the single carboxylate group evidently bridging the two Al(III).[59] The $Al_2(\mu\text{-}OH)_2^{4+}$ dimer is unstable by itself,[22, 60, 61] but becomes stabilized by bridging carboxylate,[59, 62, 63] and CO_3^{2-},[60, 61] and we suggest also by bridging phosphate in the complex $PO_4Al_2(\mu\text{-}OH)_2^+$,[34] by far the strongest of the group.

5.2. Ammonia

In aqueous solutions, hydrolysis intervenes well before Al^{3+} binds to ammonia. We provide an estimate of this useful stability constant by utilizing a graphical correlation for several metal ions of the difference of first stability constant logarithms between aspartate and malate and log K_1 for ammonia.[64] From one laboratory, the Al^{3+} stability constants logarithms are: aspartate, 7.77[65] and malate, 4.60,[66] for a difference of 3.2 log units. This difference interpolates to log $K_1 = 2.3$ for Al^{3+} and ammonia. We shall use this value in the next section.

For a ligand, L, such as ammonia that is protonated in solutions with pH < 8 to give LH, it is easy to show that the concentration ratio S of complex ML to first hydrolysis product MOH is given by

$$S = [ML]/[MOH] = K_a K_1 [LH]/K_h$$

where K_a is the acidity constant for the ligand, K_1 the stability constant of ML, and K_h the first hydrolysis constant of the metal ion, $pK_h=5.5$ in our case for Al^{3+}. In the region considered, S is independent of pH and metal ion concentration and depends only upon the three equilibrium constants and the protonated ligand concentration. If at least as much ML as MOH is required to distinguish a complex from hydrolysis products in potentiometric titrations then $S > 1$, and we rearrange the above equation to obtain $[LH] > K_h / K_aK_1 = 10^{-5.5} / 10^{-9.3} \times 10^{2.3} = 30$ M. The numerical values in the denominator are those for ammonia. The result indicates that to obtain significant complex formation between Al^{3+} and ammonia in acidic or neutral solutions, the concentration of ammonium ion needs to be an unrealistic 30 molar or greater. Due to the rapid succession of hydrolysis reactions (Section 2) and subsequent polymerizations, Al^{3+} presents an especially poor case for determining stability constants of weakly bound ligands near the borderline of extensive hydrolysis.

5.3. Amino Acids

The α-carboxylate group of amino acids is so weakly basic ($pK_1 \cong 2.4$) that meaningful interaction by itself with Al^{3+} does not occur: not even the binuclear species with carboxylate mentioned above appears.

We may provide an estimate of the stability constant for chelation of Al^{3+} by glycine by an argument similar to that given above for ammonia. A correlation has also been observed for the difference of stability constant logarithms between glycinate and glycolate and log K_1 for ammonia.[64] Now we descend deeper into the well on an old rope. From the previous section we found log K_1 = 2.3 for ammonia, and since there is no reliable stability constant for glycolate, we use that for the equivalent ligand lactate, for which the carefully determined[67] stability constant logarithm is 2.36 (at. 0.60 M ionic strength). These values interpolate to log K_1 = 5.8 for Al^{3+} and glycinate. Substituting into the same equation for [LH] in Section 5.2, we obtain $[LH] > 10^{-5.5} / 10^{-9.7} \times 10^{-5.8} = 0.025$ M. According to this estimate, chelation of Al^{3+} should become distinguishable from hydrolysis at greater than 25 mM glycine.

Two studies report results of Al^{3+} binding to simple amino acids. In the first, an insignificant influence was found by alanine on the speciation of Al^{3+}.[68] However, the highest alanine concentration used was only 12 mM, half the 25 mM required by the above estimate. In the second study, weak binding, heavily compromised by hydrolysis was found for Al^{3+} and glycine.[65] The highest glycine concentration was 20 mM and the authors estimate that for Al^{3+} and glycinate, log K_1 = 6.2. Slightly smaller stabilitiy constants are obtained for the more weakly basic serine and threonine; no hydroxy group chelation occurs with these ligands. The two studies were conducted at different temperatures, 25 and 37 °C, and ionic strengths, 0.60 M and 0.15 M. Both of these conditions favor finding the complex more easily in the latter study, as does the use of glycine which binds metal ions unusually strongly for its basicity.[69] There is a general agreement among the estimate of the previous paragraph and the two experimental studies, both at the edge of detecting complex formation. We conclude that chelation of Al^{3+} by simple α-amino acids begins to become distinguishable from hydrolysis at ligand concentrations greater than 20 mM. Future stability constant studies at up to 0.5 M amino acid should provide more precise constants.

With a less basic amino group (pK_a = 5.2) complexation becomes conveniently observable with α-picolinate (2-pyridinecarboxylate) to yield AlL^{2+} (log K_1=4.4), AlL_2^+, and $HOAlL_2$ complexes, the last from the second with pK_a=4.2.[70,71] This hydrolysis pK_a value is a highly unusual 1.3 log units more acidic than from free Al^{3+} (pK_a=5.5). We ascribe the high acidity to a reduction in coordination number from 6 to 5 either before or on deprotonation of a metal ion bound water in the bis didentate picolinate (pc) complex: $(H_2O)_2Al(pc)_2^+ \rightarrow HOAl(pc)_2 + H_3O^+$. Similar behaviour with the more basic α-amino acids would greatly complicate interpretation of their titration curves.

5.4. Enkephalin

A multinuclear magnetic resonance study with dimethyl sulfoxide as a solvent has been interpreted to indicate Al^{3+} binding to the amino group of the terminal tyrosine in the pentapeptide enkephalin.[72] However, the observed tyrosyl α-hydrogen downfield shift of 0.52 ppm on addition of Al(III) is much too large for a metalation shift and should instead be assigned to protonation, for which 0.5 ppm is a typical value.[73] There was no pH control in this study and added aquated Al(III) will behave almost as a strong acid in the non-aqueous solvent. Consistent with the discussion in the two previous sections, Al^{3+} does not bind to enkephalin in aqueous solutions.[58]

5.5 Methylmalonate

Chelation with two oxygen donors leads to stronger Al^{3+} complexes. Weak complexation occurs with the two carboxylate amino acids aspartate and glutamate.[65] Despite the greater basicity of methylmalonate,[68] it forms a weaker chelate (six-membered) with Al^{3+} than does oxalate (five-membered).[74] Methylmalonate serves as a model for metal ion binding at the two γ carboxylate groups of γ-carboxyglutamate residues in proteins, and the low stability constant log $K_1 = 5.7$ for Al^{3+} indicates that this interaction mode is not competitive with others in biological systems.

5.6. Calmodulin

Stability constants for binding to methylmalonate furnish an upper limit for Al^{3+} binding to two favorably disposed carboxylate side chains on a protein. The log $K_1 = 5.7$ value for methylmalonate and Al^{3+} supports our earlier suggestion[7,9,10] that calmodulin is unlikely to bind three Al^{3+} with stability constant values greater than 10^6. Only weak binding of Al^{3+} to calmodulin has been found by other investigators,[75,76] the latter using some of the same techniques as the early studies. With one quarter of its amino residues bearing carboxylate side chains, calmodulin is an acidic protein that should bind multiply charged metal ions as a polyelectrolyte. When it does so, magnitudes in the circular dichroism spectrum decrease in the direction of denaturation. However, it is through the calmodulin regulated proteins that contain phosphate groups that Al^{3+} will exert any influence.[9,10]

5.7. Lactate

Lactate forms weak AlL^{2+}, AlL_2^+, and AlL_3 complexes in acidic solutions where the hydroxy group retains its proton.[67] Crystals prepared at about pH 3 show a tris bidentate complex of hexacoordinate Al^{3+} for the last complex.[77] Toward more neutral solutions, AlL_2^+ undergoes loss of a lactate hydroxy group proton with $pK_a = 3.5$ to give $AlL(LH_{-1})$. This becomes the main mononuclear species even in the presence of excess lactate. No additional mononuclear species appear in this system. Slow reactions produce a variety of polynuclear species including the carboxylate dimer discussed above.[63] Administration of an AlL_3 lactate complex to a milieu near neutral pH results in decomposition of the complex.

5.8. Citrate

There is ample proof that citrate facilitates incorporation of Al(III) into mammals. Elevated Al(III) levels were found in both the brain and bones of rats fed a diet containing aluminum citrate or even just citrate.[78,79] Evidently citrate alone chelates trace Al(III) in the diet. Dosing lambs with aluminum citrate results in absorption of Al(III) and alters the balance of other minerals.[80,81] A rapidly fatal encephalopathy in four human patients with chronic renal failure has been attributed to concomitant ingestion of $Al(OH)_3$ and citrate.[82] Moreover, healthy adults taking $Al(OH)_3$ based antacids in conjunction with citric acid or citrus fruits showed substantial increases in Al(III) levels of blood[83,84] and urine.[85,86]

Citrate exists predominantly in the form of the tricarboxylate anion at pH > 6, and at 0.1 mM in the blood plasma, it is the predominate small molecule Al^{3+} binder.[7,8] A period of uncertainty in the nature of the complexes has now been resolved.[9,10,25] In neutral solutions the main species is $HOAlLH_{1}^{2-}$ followed by MLH_{1}^{-} with $pK_a \cong 6.5$ for the loss of a proton from metal ion bound water. A trinuclear species with three hydroxy group deprotonated citrates[25] has also been described in a crystal structure determination.[87] Under plasma conditions, however, only mononuclear species occur.[25] Even though much of the citrate in plasma occurs as a Ca^{2+} complex, Al^{3+} easily displaces Ca^{2+} from citrate; details appear in Section 7.2. An AlL° species with overall zero net charge exists in appreciable amounts from $3 < pH < 5$, and offers a means for citrate to act as a carrier of Al(III) across membranes.[7,8,10] However, there is a negligible flux in this system at pH 3.5 across unilamellar phospholipid vesicles.[88] It has been suggested that enhanced Al(III) absorption in the presence of citrate takes place in the proximal bowel via the paracellular pathway due to the opening of cellular tight junctions.[89]

As indicated previously[7-9] and below in Table II, citrate solubilizes Al(III) from both insoluble $Al(OH)_3$ and $AlPO_4$. Based on information from equilibrium constants, it was strongly urged in 1986 that people not take together Al(III) and citrate.[7,8] A more recent warning suggests the avoidance of citrate compounds in uremic patients.[90] However, since citrate occurs in foods as an unavoidable component of a healthy diet, it would seem more prudent instead to eliminate deliberate ingestion of Al(III) compounds. Presence or absence of citrate must be considered as a variable in all ingestion studies.

5.9. Catecholamines

With their very high $pK_a > 13$ salicylate[91] and especially catecholamines[92,93] chelate Al^{3+} between two negatively charged oxygens with high stability constants. However, their effective or conditional stability constants in neutral solutions are much reduced by competition from protons. Quantitative results are displayed in Table II and described in Section 7.1. The small amount of salicylate in the plasma cannot compete with citrate or transferrin for Al^{3+}.[9] In fluids low in citrate and transferrin the catecholamines may well become important Al^{3+} binders.[92] While DOPA and epinephrine are wholly ineffective Mg^{2+} binders at pH 7.4, they bind Al^{3+} at picomolar levels (Table II). In neutral solutions the main species is a 3 : 1 complex with the catechol moiety chelating the Al^{3+} and the ammonium group remaining protonated.[92,93] The norepinephrine-Al^{3+} complex inhibits enzymatic O-methylation but not N-methylation by catechol-O-methyl transferase.[94] This result conforms to that excepted for Al^{3+} binding only to the catechol moiety of norepinephrine. When other metal ions are deficient, Al(III) decreases catecholamine levels in the rat brain.[95] By binding to the catechol moiety of catecholamines, trace amounts of Al^{3+} may disrupt neurochemical processes.[9]

5.10. Transferrin

Transferrin is the preeminent protein carrier of Al^{3+} in the plasma.[7,96] Displacement of the 10^9 times stronger binding Fe^{3+} is unnecessary because plasma transferrin is about 50 μM in unoccupied sites. Recent experiments confirm the conclusion based on stability constants[7,96] that citrate is the low molecular weight and transferrin the high molecular weight carrier of Al(III) in rat serum.[97] We emphasize again, however, that albumin is not expected to bind Al^{3+} with anywhere near the strength of transferrin or citrate.[7,96] Experiments suggest that the transferrin receptor is transferrin and not Fe^{3+} specific, and hence Al^{3+} may gain entry to the central nervous system by the same route as does Fe^{3+}.[98] In its metal ion binding site, transferrin furnishes three anionic oxygens from two tyrosyl phenolates and one aspartyl side chains, plus one neutral nitrogen from a histidyl side chain.[99]

The transfer rate of transferrin bound Al^{3+} to excess citrate has been studied at pH 7.4, 37 °C, and 0.7 ionic strength (and evidently only an ambient HCO_3^- concentration).[100] We extrapolate the results to a plasma 0.1 mM citrate concentration to estimate a 6 hour half-life for transferrin to citrate transfer of Al^{3+}. Under physiological conditions, however, small ligands may mediate and speed the transfer.

Rat brain *ferritin* picks up Al^{3+}. Brain ferritin from Alzheimer's disease victims contained more Al(III) and Fe(III) than that from age-matched controls.[101]

5.11. Fluoride

Fluoride is an effective but mild Al^{3+} binder, and mixed hydroxo complexes occur in neutral solutions.[102] A multinuclear magnetic resonance study has shown that a mixed 1:1:1 complex of F^-, Al^{3+}, and ADP occurs in the statistically expected amount based on stability constants of the 2:1 binary complexes.[47] The maximum number of F^- occuring in a ternary complex is three, $(ADP)AlF_3^{3-}$. Because Al^{3+} binds exclusively at the phosphates, similar results are expected for other nucleoside diphosphates such as GDP. It is likely that Al(III) remains hexacoordinate in all ternary complexes with phosphates including those with protein.

Owing to the insolubility of CaF_2, there is an incompatibility among the usual Al(III), F^-, and Ca^{2+} concentrations in many experiments with proteins, especially G-proteins. The solubility product concentration constant for CaF_2 is 0.3 $mM^3 = K_{sp} = [Ca^{2+}][F^-]^2$.[9] A typical 5 mM F^- ambient concentration permits only 12 μM Ca^{2+}. Maintenance of the Ca^{2+} concentration at a higher level excludes the AlF_4^- complex. Mg^{2+} also combines with F^- but more weakly than Ca^{2+}.[9,21] About half the total F^- in seawater is bound as MgF^+.

6. pAl = −Log [Al^{3+}]

Results obtained above for Al^{3+} hydrolysis, inorganic phosphate precipitation, and complexation to a variety of ligands are summarized as the negative logarithm of the free, aqueous Al^{3+} concentration, pAl=− log [Al^{3+}] in Table II and figure 2. Table II lists pAl values at pH 6.6 and 7.4 for several systems under the conditions indicated in the Table. Weak Al^{3+} binders are at the top and strong binders at the bottom of Table II. The values are derived by applying equation 10 (below) to the soluble ligand systems. The increasing pAl values as one goes down the Table indicate a decreasing free Al^{3+} concentration. Thus, since 0.1 mM citrate lies lower in Table II then does 1 mM ATP^{4-}, we predict that citrate will withdraw Al^{3+} from ATP^{4-}, and experimentally citrate has been used for this purpose.[103] Despite high normal stability constants as indicated in the second column, salicylate[91] and catecholamines[92] bind Al^{3+} relatively weakly because of competition from the proton in the strongly basic ligands leading to the low conditional stability constants listed in the third and fifth columns.

Figure 2 shows for several ligands pAl as a function of pH. Stability constant results for maltol,[104] nitrilotriacetate (NTA),[105] citrate,[9,10,25] EDTA,[106] and deferoxamine (dfo),[107] represented by the solid curves have been processed over the pH range according to equation 10 (below). For these five ligands a 1 μM total Al(III) is assumed as indicated by the left pAl=6 intercept of the curves on the ordinate at low pH where in all cases the aluminum appears unbound as Al^{3+}. As the pH increases, strong differentiation in binding strength occurs with maltol the weakest and deferoxamine the strongest Al^{3+} binder. Kaolinite is the least soluble aluminum silicate.[108] The rectangle at pH 7.4 labeled Tf represents a range of reported values for transferrin under blood plasma conditions.[96,109]

Though it is slightly more basic than maltol, acetylacetone (acac or 2,4-pentanedione) exhibits correspondingly greater stability constants in up to a 3:1 complex[110] so that the two ligands yield similar conditional stability constants in neutral solutions. Thus a curve for acac closely tracks that shown for maltol in figure 2. (However, the acac stability constants for Al^{3+} were determined before the advent of modern methods and should be checked.) As indicated in figure 2, at 1 μM total Al(III) the 3:1 maltol complex is unstable with respect to Al(III) hydrolysis products. At 1 mM total Al(III), the 3:1 maltol complex dominates from 5 < pH < 9.3, where $Al(OH)_4^-$ becomes dominant. Therefore, the concentrations at which maltol and acac complexes are dosed may be important and may account for some inconsistent results among different investigations. The acac complex shows stronger effects than that of maltol in inducing osmotic fragility and acanthocyte formation in erythrocytes. These differences have been attributed to the greater lipophilicity of the 3 : 1 acac complex.[111]

The dashed curve labeled $Al(OH)_4^-$ near pH 9 represents the pAl of 1 μM total Al(III) allowed by the hydrolysis equilibria 1 – 4 depicted in the upper half of figure 1. The left hand ordinate intercept of the dashed curve is also at pAl=6 where the predominant species is Al^{3+}. As indicated by the distribution curves in the upper half of figure 1, the species represented by the dashed curve in figure 2 pass through all the species of equations 1 – 4 in the region $5 < pH < 7$.

The dashed curve and the dotted line labeled $Al(OH)_3(s)$ apply to all aqueous solutions containing Al(III). The dotted line of slope 3 represents the pAl allowed by the solubility of an amorphous $Al(OH)_3$ according to equation 5. The dotted curve labeled Pi shows the pAl permitted by the solubility of $AlPO_4$ in the presence of 4.5 mM total phosphate according to equation 6. At about pH 8.3 the free Al^{3+} permitted by 4.5 mM inorganic phosphate exceeds that allowed by the solubility of an amorphous $Al(OH)_3$. The two dotted curves in figure 2 are not limited by the 1 μM total Al(III) concentration assumed for all the other curves.

7. Conditional Stability Constants

7.1. Ligand Protonation

In addition to the hydrolysis features already considered, the amount of free, aqueous Al^{3+} in solution depends upon several variables: ligands present, their stability constants with Al^{3+}, and total Al(III) to total ligand mole ratio. For ligands with protons competing with metal ion for binding sites in the pH range of interest, the pH is also a variable. Thus, instead of simple association of metal ion and basic ligand, $Al^{3+} + L \rightarrow AlL$, the relevant reaction may become displacement of a proton from the acidic ligand by the metal ion, $Al^{3+} + HL \rightarrow AlL + H^+$. For ligands containing amino, phenolate, and catecholate groups, the amount of free, aqueous Al^{3+} in neutral solutions becomes pH dependent. Thus for these ligands, listed stability constants overstate effective binding strengths and need to be reduced in magnitude to reflect competition of the proton with the metal ion for basic binding sites. The most practical method to allow for proton-metal ion competition at a ligand is to calculate *conditional* stability constants applicable to a single pH. Conditional stability constants may also allow for deprotonation of metal ion coordinated water that yields a pH dependent stability in some Al^{3+} complexes such as with citrate, nitrilotriacetate, and EDTA.

We have used nitrilotriacetate (NTA) to set the free, aqueous Al^{3+} in a study of the polymerization of tubulin into microtubules.[112] The standard stability constant refers to the reaction of basic ligand

$$Al^{3+} + L^{3-} \rightleftarrows AlL^{\circ}$$

$$K_s = [AlL^{\circ}] / [Al^{3+}][L^{3-}]$$

where $\log K_s = 11.1$.[105] In neutral solutions, however, owing to the ammonium group with $pK_a = 8.9$, most of the ligand appears as a species with 2– net charge.

$$HL^{2-} \rightleftarrows H^+ + L^{3-}$$

$$K_a = (H^+)[L^{3-}] / [HL^{2-}]$$

where the parentheses represent activity and the brackets concentration. Thus in neutral solutions we have the displacement reaction

$$Al^{3+} + HL^{2-} \rightleftarrows H^+ + AlL^{\circ}$$

$$K'_c = [AlL^{\circ}] / [Al^{3+}][HL^{2-}]$$

where K'_c is defined without inclusion of (H^+). (Editorial errors made in proof result in incorrect equations in Section 13 of reference 9.) The concentration of deprotonated ligand L^{3-} available to the metal ion is reduced by occurrence of protonated species. The fraction of unbound, depronated ligand is given by $\alpha = K_a / [(H^+) + K_a]$ so that $0 < \alpha < 1$. The conditional, pH dependent stability constant is given by $K'_c = \alpha K_s$ or

$$\log K'_c = \log K_s - pK_a - \log [(H^+) + K_a] \quad (7)$$

In the limit where pH $\gg$ pK_a, the basic form of unbound ligand predominates and log K'_c= log K_s. For pH $\ll$ pK_a unbound ligand is predominantly protonated and log K'_c= log K_s − pK_a + pH, and the conditional constant displays a pH dependence. When the pH is within 1.5 log units of pK_a, the complete equation 7 should be employed. For NTA at pH 7.4 we have log K'_c=11.1−8.9+7.4=9.6. The conditional stability constant provides a measure of complex stability under specific pH conditions where there is a protonated ligand. By allowing for withdrawal of deprotonated ligand from solution by protonation, the value of the conditional stability constant K'_c becomes less than the standard stability constant K_s.

For EDTA or catecholates where Al^{3+} displaces two protons, the fraction of unbound, doubly deprotonated ligand becomes $\alpha = K_{a1} \cdot K_{a2} / [(H^+)^2 + (H^+)K_{a1} + K_{a1}K_{a2}]$, where $pK_{a1} < pK_{a2}$. The definition of α may be extended to three or more basic groups. In all cases the relation between the conditional and standard stability constants is given by $K'_c = \alpha\ K_s$.

7.2. Complex Deprotonation

A crystal structure determination of the NTA-Al(III) complex shows a quadridentate ligand bound via three carboxylate and one amino groups with two water molecules completing the ligands bound to hexacoordinate Al^{3+}.[113] One of the bound waters undergoes deprotonation with pK_b=5.4[105] according to

$$(H_2O)_2AlL^\circ \rightleftarrows HOAlL^- + H_3O^+$$

$$K_b = (H^+)[HOAlL^-] / [(H_2O)_2AlL^\circ]$$

(We account for the low pK_b=5.4 by a reduction in coordination number with a loss of a water molecule to yield a pentacoordinate Al(III) on the right hand side.[9,22] The subsequent «deprotonation» with a relatively high pK_{b2}=8.2 then refers to addition of OH^-.) The deprotonation reaction effectively strengthens the interaction between metal ion and ligand, increases the effective stability constant, and decreases the free, aqueous Al^{3+} concentration. The fraction of non-deprotonated complex is given by $\beta = (H^+) / [(H^+) + K_b]$ so that $0 < \beta < 1$. The conditional, pH dependent stability constant is now given by $K_c = \alpha K_s / \beta$ or

$$\log K_c = \log \alpha + \log K_s + pH + \log [(H^+) + K_b] \tag{8}$$

where $\log \alpha = -pK_a - \log [(H^+) + K_a]$.

For NTA at pH 7.4 the conditional stability constant K_c refers mainly to the reaction

$$(H_2O)Al^{3+} + HL^{2-} \rightleftarrows HOAlL^- + 2H^+$$

$$K_c = [HOAlL^-] / ([Al^{3+}][HL^{2-}])$$

We now have log K_c=11.1−8.9+7.4+7.4−5.4=11.6. Note that at pH 7.4, the stability promoting effect of the deprotonated complex more than offsets (by 0.5 log units) the destabilizing effect of an appreciable fraction of monopronated unbound ligand.

The ideas developed here apply to other systems as well. The basic equation is $K_c = \alpha K_s / \beta$ with both $0 < \alpha, \beta < 1$. If more than one protonation of the free ligand occurs, or if more than one deprotonation from a complex takes place (as for citrate), the expressions for α and β, respectively, need to be generalized from the monoproton case of NTA considered as a prototype in this section.

For the citric acid complex of Al^{3+} two protons are lost by pH 7.4, the first from the citrate ligand with pK_{1b}=3.7 and the second from an Al^{3+}-bound water with pK_{2b}=6.5.[9,25] At pH 7.4 the main reaction occurring is

$$(H_2O)Al^{3+} + Cit^{3-} \rightleftarrows HOAl(CitH_{-1})^{2-} + 2H^+$$

with the conditional constant

$$K_{7.4} = [HOAl(CitH_{-1})^{2-}] / [Al^{3+}][Cit^{3-}]$$

The expression for the fraction of nondepronated complex becomes

$$\beta = (H^+)^2 / [(H^+)^2 + (H^+)K_{1b} + k_{1b}K_{2b}]$$

and at pH 7.4, $\beta = 2.2 \times 10^{-5}$. Since the third deprotonation from the free ligand occurs with pK_3=5.7, at pH 7.4 we calculate that α=0.99. From log K_c = log K_s + log α − log β we now obtain for the conditional stability constant at pH 7.4, log $K_{7.4}$=8.1−0.0+4.6=12.7. Alternatively, since at pH 7.4 we have pK_{1b}, pK_{2b}, & $pK_3 \ll$ 7.4, the expression for the conditional constant reduces to log $K_{7.4} \cong$ log K_s+2 pH−

$pK_{1b} - pK_{2b} = 8.1 + 2 \times 7.4 - 3.7 - 6.5 = 12.7$. In the case of the citrate-Al^{3+} complex at pH 7.4, the effective stability constant is 4×10^4 times greater than the tabulated stability constant owing mainly to deprotonation of the complex in quite acidic solutions. These values appear in the 13th row of Table II.

Even though most citrate in the plasma is complexed by Ca^{2+} or Mg^{2+}, Al^{3+} still easily displaces the alkaline earth metal ions. They form simple equimolar complexes in neutral solutions with the citrate stability constants for Ca^{2+} ($10^{3.5}$) and Mg^{2+} ($10^{3.4}$) nearly equal,[114] so we utilize the Ca^{2+} value ($K_{Ca} = 10^{3.5}$) and take the sum of the plasma concentrations of the two metal ions as 3 mM. The relevant displacement reaction in neutral solutions is

$$Al^{3+} + CaCit^- \rightleftarrows HOAl(CitH_{-1})^{2-} + 2H^+ + Ca^{2+}$$

with the conditional equilibrium constant

$$K_c = [HOAl(CitH_{-1})^{2-}][Ca^{2+}] / [Al^{3+}][CaCit^-] = K_{7.4} / K_{Ca} = 10^{9.3} \quad (9)$$

where $K_{7.4}$ is given in the paragraph above. The effect of the alkaline earth metal ion binding to citrate is to reduce the conditional stability constant by 3.5 log units.

The ratio of citrate complexed to free Al^{3+} in the presence of the alkaline earth metal ions is found by rearranging eq 9 and substituting the relevant values of 0.1 mM complexed citrate and 3 mM dispositive ions to obtain

$$[HOAl(CitH_{-1})^{2-}] / [Al^{3+}] = K_c\ [CaCit^{3-}] / [Ca^{2+}] \cong 10^{9.3} \times 10^{-4} / 3 \times 10^{-3} = 10^{7.8}$$

Including consideration of alkaline earth cations in the plasma, there is an almost 10^8 mole ratio of citrate bound to unbound Al^{3+}.

In view of the displacement of Ca^{2+} from citrate by Al^{3+}, the advice to give calcium citrate to uremic patients[115] requires that ingestion of Al(III) be scrupulously avoided. Other Ca^{2+} salts might be used instead. Calcium acetate has been suggested, and it has the further advantage that it works more rapidly than calcium citrate.[116]

7.3. Metal Ion Buffers

Only rarely in experiments with Al(III) has the metal ion concentration been controlled reliably at known values.[112] Metal ion buffers are analogous to pH buffers except that free metal ion concentration is controlled in the presence of excess ligand. As with protonic equilibria, the most effective buffering occurs when the complexed ligand to total ligand molar concentrations lie between 0.15 to 0.85. Since ligand concentration exceeds that of tightly bound metal ion, the total metal ion to total ligand molar ratio R should also lie between 0.15 to 0.85.

The development of the previous section in terms of the conditional stability constant K_c leads to the equation for metal ion buffering. For the NTA–Al^{3+} system of equation 8 the total ligand concentration $C_L = [HL^{2-}] + [HOAlL^-]$ and the total Al(III) concentration $C_M = [HOAlL^-]$ as ligand occurs in excess, and the metal ion is tightly bound so that for the purposes of addition and subtraction the concentration of free Al^{3+} is negligible. Substituting in the expression for the conditional stability constant, rearranging, and noting $R = C_M / C_L$, we obtain

$$[Al^{3+}] = R / [(1-R)K_c]$$
$$\text{and } pAl = \log K_c + \log [(1-R)/R] \quad (10)$$

This result is general. The free metal ion concentration is given by equation 10 where the conditional stability constant applies to the specific pH for which it was calculated. For systems formally analogous to NTA, K_c is given by

$$K_c = K_s [1 + K_b / (H^+)] / [1 + (H^+) / K_a] \quad (11)$$

The $(H^+) / K_a$ term accounts for protonated ligand and the $K_b / (H^+)$ term for the deprotonation of Al^{3+}-bound water in the complex.

The important metal ion buffer equations 10 & 11 express the free metal ion concentration as a function of known equilibrium constants, pH, and only the ratio R of total metal ion to excess total ligand molar concentrations, not their absolute concentrations. To vary the free metal ion concentration at a fixed pH, it is necessary to vary R. At a fixed pH the right-hand side of equation 10 describes a sigmoidal curve symmetrical about $R = 0.5$, similar to a titration curve. We have for the midpoint at $R =$

0.5, $[Al^{3+}] = 1/K_c$. At pH 7.4 and R=0.5 we obtain for NTA, pAl=−log $[Al^{3+}]$=11.6 and for citrate, pAl=12.7. Both metal ion concentrations are very low but controlled. For the same R value the pAl for the two buffer systems differs by only 0.8 units or a factor of 6 in $[Al^{3+}]$. Experiments with the two different kinds of ligands should serve as a check on one another. However, due to time-dependent reactions in citrate-Al^{3+} solutions, this buffer system should be used carefully. Both buffer systems were used in a recent evaluation of the stability constant for Al^{3+} binding to the serum protein transferrin.[96] Investigation of the polymerization of tubulin to microtubules used the NTA buffer system to control less than nanomolar concentrations of Al^{3+}.[112]

It is reasonable to vary R from 0.15 to 0.85 to obtain a range of 1.5 log units in pAl. For a wider range, vary pH or find another ligand for a new metal ion buffer system. The condition where R=1 or $C_M = C_L$ should be avoided as the solution is unbuffered with wide swings in the value of pAl for even small deviations from exact equality (analogous to the end point in a tiration curve).

With allowance for an additional ammonium group deprotonation in the free ligand, the final equation for EDTA complexation resembles that for NTA with deprotonation from the complex. Substitution of the EDTA stability constants, pH 7.4, pK_b=5.9, and R= 0.5 yields at the midpoint of the metal ion buffer range, pAl=16.4. The EDTA free Al^{3+} concentration is 1/1,200 that from NTA. However, equilibrium in the EDTA complex of Al^{3+} is reached slowly, and experiments involving it need careful monitoring. The EGTA complex of Al^{3+} is poorly defined and should not be used as an Al^{3+} buffer.

References

1. Hydes D.: *Nature*, **268**, 136, 1977.
2. Hydes D.: *Science*, **205**, 1260, 1979.
3. Stoffyn M.: *Science*, **203**, 651, 1979.
4. Jorgensen S.E., Jensen A.: *Metal Ions Biol. Syst.*, **18**, 61, 1984.
5. Matsumoto H., Hirasawa F., Morimura S., Takahasi E.: *Plant Cell Physiol.*, **17**, 627, 1976.
6. Koch K.R., Pougnet M.A.B., de Villiers S., Monteagudo F.: *Nature*, **333**, 122, 1988.
7. Martin R.B.: *Clin. Chem.*, **32**, 1797, 1986.
8. Martin R.B.: *J. Inorg. Biochem.*, **28**, 181, 1986.
9. Martin R.B.: *Metal Ions Biol. Syst.*, **24**, 1, 1988.
10. Martin R.B.: in *Aluminium and Renal Failure*, deBroe M.E., Coburn J.W., Eds. (Kluwer Academic Publishers, Dordrecht, 1990) pp. 7- 26.
11. Dennis E.J.: *Fert. Solutions*, **15**, 44-54, 1971.
12. Fontenot J.P., Allen V.G., Bunce G.E., Goff J.P.: *J. Anim. Sci.*, **67**, 3445, 1989.
13. Wills M.R., Savory J.: *Lancet*, **ii**, 29, 1983.
14. Hewitt C.D., Savory J., Wills M.R.: *Clinics Lab. Med.*, **10**, 403, 1990.
15. Chazan J.A., Lew N.L., Lowrie E.G.: *Arch. Intern. Med.*, **151**, 319, 1991.
16. Eastwood J.R., Levin G.E., Pazianas M., Taylor A.P., Denton J., Freemont A.J.: *Lancet*, **ii**, 462, 1990.
17. Mozar H.N., Bal D.G., Howard J.T.: *J. Am. Med. Assn.*, **257**, 1503, 1987.
18. Corain B., Bombi G.G., Tapparo A., Nicolini M., Zatta P., Perazzolo M., Favarato M.: *Environ. Health Perspect.*, **89**, 233, 1990.
19. Martin R.B.: *Metal Ions Biol. Syst.*, **20**, 21, 1986.
20. Macdonald T.L., Martin R.B.: *Trends Biochem. Sci.*, **13**, 15, 1988.
21. Martin R.B.: *Metal Ions Biol. Syst.*, **26**, 1, 1990.
22. Martin R.B.: *J. Inorg. Biochem.*, **43**, in press, 1991.
23. Hem J.D.: *Kidney Internat.*, **29**, *Suppl.* 18, S3, 1986.
24. Bates R.G.: *Pure and Appl. Chem.*, **57**, 531, 1985.
25. Ohman L.: *Inorg. Chem.*, **27**, 2565, 1988.
26. Ohman L.: *Inorg. Chem.*, **28**, 3629, 1989.
27. Akeson M.A., Munns D.N., Burau R.G.: *Biochim. Biophys. Acta*, **986**, 33, 1989.
28. Kiss T., Sovago I., Martin R.B.: *Inorg. Chem.*, **30**, 2130, 1991.
29. Sovago I., Kiss T., Martin R.B.: *Polyhedron*, **9**, 189, 1990.
30. Abreo K., Brown S.T., Sella M.: *Amer. J. Kidney Dis.*, **13**, 465, 1989.
31. Sigel H., Tribolet R., Malini-Balakrishnan R., Martin R.B.: *Inorg. Chem.*, **26**, 2149, 1987.
32. Massoud S.S., Sigel H.: *Inorg. Chem.*, **27**, 1447, 1988.
33. Duffield J.R., Edwards K., Evans D.A., Morrish D.M., Vobe R.A., Williams D.R.: *J. Coord. Chem.*, **23**, 277, 1991.
34. Dayde S., Fiella M., Berthon G.: *J. Inorg. Biochem.*, **38**, 241, 1990.
35. Perl D.P., Brody A.R.: *Science*, **208**, 297-299, 1980.
36. Perl D., Gajdusek D.C., Garruto R.M., Yanagihara R.T., Gibbs C.J.: *Science*, **217**, 1053, 1982.
37. Crapper D.R., Quittkat S., Krishnan S.S., Dalton A.J., DeBoni U.: *Neuropathol.*, **50**, 19, 1980.

38. DeBoni U., Scott J.W., Crapper D.R.: *Histochem.*, **40**, 31, 1974.
39. Wen G.Y., Wisniewski H.M.: *Neuropathol.*, **68**, 175, 1985.
40. Karlik S.J., Eichhorn G.L., Lewis P.N., Crapper D.R.: *Biochemistry*, **19**, 5991, 1980.
41. Karlik S.J., Eichhorn G.L.: *J. Inorg. Biochem.*, **37**, 259, 1989.
42. Karlik S.J., Chong A.A., Eichhorn G.L., DeBoni U.: *Neurotoxicology*, **10**, 167, 1989.
43. Dryssen D., Haraldsson C., Nyberg E., Wedborg W.: *J. Inorg. Biochem.*, **29**, 67, 1987.
44. Sternberger N.H., Sternberger L.A., Ulrich J.: *Proc. Natl. Acad. Sci.*, **82**, 4274, 1985.
45. Grundke-Iqbal, Iqbal K., Tung Y., Quinlan M., Wisniewski H.M., Binder L.I.: *Proc. Natl. Acad. Sci.*, **83**, 4913, 1986.
46. Diaz-Nido J., Avila J.: *Neuroscience Lett.*, **110**, 221, 1990.
47. Nelson D.J., Martin R.B.: *J. Inorg. Biochem.*, **43**, 37, 1991.
48. Martin R.B.: *Metal Ions Biol. Syst.*, **17**, 1, 1984.
49. Dunstan C.R., Hills E., Norman A.W.: *Quart. J. Med.*, **55**, 127, 1985.
50. Hugi-Cleary D., Helm L., Merbach A.E.: *Helv. Chim. Acta*, **68**, 545, 1985.
51. Kido H., Saito K.: *J. Am. Chem. Soc.*, **110**, 3187, 1988.
52. Crumbliss A.L., Garrison J.M.: *Comments Inorg. Chem.*, **8**, 1, 1988.
53. Secco F., Venturini M.: *Inorg. Chem.*, **14**, 1978, 1975.
54. Perlmutter-Hayman B., Tapuhi E.: *Inorg. Chem.*, **16**, 2742, 1977.
55. Garrison J.M., Crumbliss A.L.: *Inorg. Chim. Acta*, **138**, 61, 1987.
56. Clevette D.J., Orvig C.: *Polyhedron*, **9**, 151, 1990.
57. Martin R.B.: *Metal Ions Biol. Syst.*, **8**, 57, 1979.
58. Putterill J.J., Gardner R.C.: *Biochim. Biophys. Acta*, **964**, 137, 1988.
59. Marklund E., Ohman L., Sjoberg S.: *Acta Chem. Scand.*, **43**, 641, 1989.
60. Ohman L., Forsling W.: *Acta Chem. Scand.*, **A35**, 795, 1981.
61. Hedlund T., Sjoberg S., Ohman L.: *Acta Chem. Scand.*, **A41**, 197, 1987.
62. Ohman L.: *Acta Chem. Scand.*, **45**, 258, 1991.
63. Marklund E., Ohman L.: *Acta Chem. Scand.*, **44**, 228, 1990.
64. Mulla F., Marsicano F., Nakani B.S., Hancock R.D.: *Inorg. Chem.*, **24**, 3076, 1985.
65. Dayde S.: *Universite de Toulose*, Thesis, 1990.
66. Venturini M., Berthon G., May P.M.: *Rec. Trav. Chim.*, **106**, 406, 1987.
67. Marklund E., Sjoberg S., Ohman L.: *Acta Chem. Scand.*, **A40**, 367, 1986.
68. Marklund E., Ohman L.: *Acta Chem. Scand.*, **44**, 353, 1990.
69. Martin R.B.: *Metal Ions Biol. Syst.*, **9**, 1, 1979.
70. Jons O., Johansen E.S.: *Inorg. Chim. Acta*, **151**, 129, 1988.
71. Findlow J.A., Duffield J.R., Williams D.R.: *Chem. Speciation. Bioavail.*, **2**, 3, 1990.
72. Mazarguil H., Haran R., Laussac J.: *Biochim. Biophys. Acta*, **717**, 465, 1982.
73. Mathur R., Martin R.B.: *J. Phys. Chem.*, **69**, 668, 1965.
74. Sjoberg S., Ohman L.: *J. Chem. Soc. Dalton*, 2665, 1985.
75. Richardt G., Federolf G., Habermann E.: *Arch. Toxicol.*, **57**, 257, 1985.
76. You G., Nelson D.J.: *J. Inorg. Biochem.*, **41**, 283, 1991.
77. Bombi G.G., Corain B., Sheikh-Osman A.A., Valle G.C.: *Inorg. Chimica Acta*, **171**, 79, 1990.
78. Slanina P., Falkeborn Y., Frech W., Cedergren A.: *Fd. Chem. Toxic.*, **22**, 391, 1984.
79. Slanina P., Frech W., Bernhardson A., Cedergren A., Mattsson P.: *Acta Pharmacol. Toxicol.*, **56**, 331, 1985.
80. Allen V.G., Fontenot J.P., Rahnema S.H.: *J. Anim. Sci.*, **68**, 2496, 1990.
81. Allen V.G., Fontenot J.P., Rahnema S.H.: *J. Anim. Sci.*, **69**, 792, 1991.
82. Bakir A.A., Hryhorczuk D.O., Berman E., Dunea G.: *Trans. Am. Soc. Artif. Intern. Oragans*, **32**, 171, 1986.
83. Slanina P., Frech W., Ekstrom L., Loof L., Slorach S., Cedergren A.: *Clin. Chem.*, **32**, 539, 1986.
84. Weberg R., Berstad A.: *Europ. J. Clin. Invest.*, **16**, 428, 1986.
85. Bakir A.A., Hryhorczuk D.O., Ahmed S., Hessl S.M., Levy P.S., Spengler R., Dunea G.: *Clin. Nephrology*, **31**, 40, 1989.
86. Walker J.A., Sherman R.A., Cody R.P.: *Arch. Intern. Med.*, **150**, 2037, 1990.
87. Feng T.L., Gurian P.L., Healy M.D., Barron A.R.: *Inorg. Chem.*, **29**, 408, 1990.
88. Akeson M.A., Munns D.N.: *Biochim. Biophys. Acta*, **984**, 200, 1989.
89. Froment D.P., Molitoris B.A., Buddington B., Miller N., Alfrey A.C.: *Kidney Int.*, **36**, 978, 1989.
90. Molitoris B.A., Froment D.H., Mackenzie T.A, Huffer W.H., Alfrey A.C.: *Kidney Int.*, **36**, 949, 1989.
91. Ohman L., Sjoberg S.: *Acta Chem. Scand*, **A37**, 875, 1983.
92. Kiss T., Sovago I., Martin R.B.: *J. Am. Chem. Soc.*, **111**, 3611, 1989.
93. Kiss T., Sovago I., Martin R.B.: *Polyhedron*, **10**, 1401, 1991.
94. Mason L., Weinkove C.: *Ann. Clin. Biochem.*, **20**, 105, 1983.
95. Wenk G.L., Stemmer K.L.: *Neurotoxicology*, **2**, 347, 1981.
96. Martin R.B., Savory J., Brown S., Bertholf R.L., Wills M.: *Clin. Chem.*, **33**, 405, 1987.
97. vanGinkel M.F., vanderVoet G.B., vanEijk H.G., deWolff F.A.: *J. Clin. Chem. Clin. Biochem.*, **28**, 459, 1990.

98. Roskams A.J., Connor J.R.: *Proc. Natl. Acad. Sci.*, **87**, 9024, 1990.
99. Baker E.N., Rumball S.V., Anderson B.F.: *TIBS*, **12**, 350-353, 1987.
100. Marques H.M.: *J. Inorg. Biochem.*, **41**, 187, 1991.
101. Fleming J.T., Joshi J.G.: *Proc. Natl. Acad. Sci.*, **84**, 7866, 1987.
102. Martin R.B.: *Biochem. Biophys. Res. Commun.*, **155**, 1194, 1988.
103. Womack F.C., Colowick S.P.: *Proc. Natl. Acad. Sci.*, **76**, 5080, 1979.
104. Hedlung T., Ohman L.: *Acta Chem. Scand.*, **A42**, 702, 1988.
105. Ohman L.: *Polyhedron*, **9**, 199, 1990.
106. Bhat T.R., Das R.R., Shankar J.: *Indian J. Chem.*, **5**, 324, 1967.
107. Evers A., Hancock R.D., Martell A.E., Motekaitis R.J.: *Inorg. Chem.*, **28**, 2189, 1989.
108. Martin R.B.: *Polyhedron*, **9**, 193, 1990.
109. Harris W.R., Sheldon J.: *Inorg. Chem.*, **29**, 119, 1990.
110. Izatt R.M., Fernelius W.C., Haas J.C.G., Block B.P.: *J. Phys. Chem.*, **59**, 170, 1955.
111. Zatta P., Perazzolo M., Corain B.: *Toxicology Lett.*, **45**, 15, 1989.
112. MacDonald T.L., Humphreys W.G., Martin R.B.: *Science*, **236**, 183, 1987.
113. Valle G.C., Bombi G.G., Corain B., Favarato M., Zatta P.: *J. Chem Soc. Dalton Trans.*, 1513, 1989.
114. Field T.B., Coburn J., McCourt J.L., McBryde W.A.E.: *Anal. Chim. Acta*, **74**, 101, 1975.
115. Cushner H.M., Copley J.B. Lindberg J.S., Foulks C.J.: *Kidney Int.*, **33**, 95, 1988.
116. Sheikh M.S., Maguire J.A., Emmett M., Santana C.A., Nicar M.J., Schiller L.R., Fordtran J.S.: *J. Clin. Invest.*, **83**, 66, 1989.

Aluminium Determination in Complex Matrixes

S. Costantini, R. Giordano

In the last few years numerous scientific studies have brought to light the potential toxicity of aluminium, giving particular reference to uremic subjects undergoing long-term dialysis. The presence of the element in the haematic flow of these patients sometimes induces serious neurological and osteodystrophyc complications, forming a particularly unfavourable clinical picture.[1-3]

Even though toxic effects have primarily been found within the ambit of dialysis, numerous findings indicate that aluminium may also exert its negative action in people who do not have any renal damage. In fact, the element seems to be implicated in the pathogenesis of Alzheimer's disease, but its precise role has not yet been explained by current evidence. Only further research can clarify this.[4,5]

Aluminium is present in very small amounts in living organisms but is abundant in the environment. Composing 8% of the eart's crust, the element is the most abundant metal and the third most abundant of all elements after oxygen and silicon. Despite all this, it is usually excluded from normal biochemical and metabolic processes because of its chemical nature. This is mainly due to the low solubility of its chemical forms, namely silicates, phosphates and oxides, which renders the aluminium chemically unavailable. The only accessible oxidation state in biological systems is +3, since the standard reduction potential for the reaction $Al^{+3} + 3e^- \rightleftarrows Al^o$ is highly negative (− 1.67 V). Its high electric charge and the small ionic radius (0.51 Å) give Al^{+3} a strong polarizing effect on adjacent atoms. Therefore, the element is too reactive to be found free in nature. Al^{+3} has never been demonstrated as having a definite biological function, suggesting that the ion possesses properties which are incompatible with fundamental life processes.[6,7]

Aluminium concentrations in natural waters normally are small because the element tends to form stable structures with hydroxides that precipitate from solution. Concentration levels in natural fresh waters vary from a few tens to some hundreds of micrograms per liter, whereas the concentration of aluminium dissolved in sea water is generally lower than 1 μ g/L.[8] Aluminium is present in air in the form of aluminosilicates associated with dust particles; this is the result of the weathering of aluminosilicate rocks. Concentrations have been estimated at 50-500 ng/m^3 in rural areas and 100-5000 ng/m^3 in urban environments. Not only is Al ubiquitous, but further problems arise from its wide use in industry, resulting in possible intake on the part of man through numerous routes.[9]

In the light of the knowledge that aluminium plays an important role in the agronomical filed, particularly as a material used by the industry in preparing and transforming foods, its role as a food contaminant must be considered attentively, along with the potential risks for consumers. A recent example was brought to light by the presence of aluminium in infant formulae. In this case, the concentration of the element depends both on the raw materials used and the technological procedures employed. In particular, calcium hydroxide used for preparing demineralized milk sera can contain high amounts of aluminium. The concentrations in breast milk vary from about 0.004 to 1 mg/kg (European range 0.02 - 0.1 mg/kg), whereas the Al content in cow's milk is in the range of 0.02 - 0.9 mg/kg (European mean 0.1 mg/kg). These values are greatly exceeded in soya-based formulae.[10, 11] The dietary intake of aluminium can vary from 3 to 30 mg/day, but a reasonable average dietary intake is problably about 5 mg/day.

In plants, aluminium toxicity seems to act at

the growth level and its accumulation is a function of the acidity of the soil which influences its bioavailability. The few studies carried out in this specific field of application suggest that plant tolerance to aluminium is due to the combination of two factors (exclusion mechanisms and internal tolerance) through the formation of a pH barrier induced in the rhizosphere, and also the action of radical exudates which have chelating properties.

Many analytical techniques are available for measuring aluminium, such as X-ray fluorescence, spectrofluorimetric analysis, neutron activation, atomic emission spectroscopy and atomic absorption spectrometry. Nevertheless, interference from other metals and/or inadequate sensitivity render most of the techniques unsuitable for aluminium determination in complex matrixes. Recently, the lack of energy in flame emission spectroscopy has been overcome by the use of inductively coupled plasma (ICP) as an excitation source. However, the intense broad emission of calcium produced by argon-plasma increases the aluminium background and raises the detection limit of this element.[12] Owing to the wide analytical possibilities offered by the present instrumental solutions, atomic absorption spectrometry is currently the method of choice for laboratories measuring aluminium. However, the potentiality of the flame technique, even with the hotter nitrous-oxide acetylene flame, is sometimes restricted by sensitivity problems since it is difficult to measure the element correctly at concentrations lower than 1 mg per liter. On the contrary, the high analytical sensitivity offered by the electrothermal graphite furnace technique make it possible to determine aluminium at very low concentrations with very small sample sizes.

In this technique a small volume of sample, usually ranging from 5 to 20 μL, is placed in a graphite tube, inserted into the furnace and directly heated by the current. The tube is mounted in the path of a beam which is produced by a hollow cathode lamp whose cathode is constructed in the metal being analyzed, in this case aluminium (Fig. 1). In the furnace, after a thermic pre-treatment, the sample is quickly atomized so that the atoms of the analyte produced by the process are able to absorb the radiation at the resonance wavelength. The initial light intensity is then decreased by an amount determined by the atom concentrations. Absorbance is the most convenient term for characterizing light absorption, since this quantity has a linear relationship with concentration, in agreement with Beer's Law: $A = abc$ (A = absorbance; a = absorption coefficient which is characteristic of the absorbing species; b = length of light path intercepted by the absorption cell; c = concentration of the absorbing species in the absorption cell). In this way, it is possible to obtain an absorption signal of short duration, whose peak increases along with the increasing of the atom concentration in the furnace, returning to zero when the atoms diffuse in the graphite tube. Depending on the instrumental conditions, either the peak height or the peak area are directly proportional to the concentration of the element being examined.

There are a number of potential pathways by which gas phase atoms may be formed at the atomization stage.[13] Several studies performed on graphite furnace kinetics (Table I)

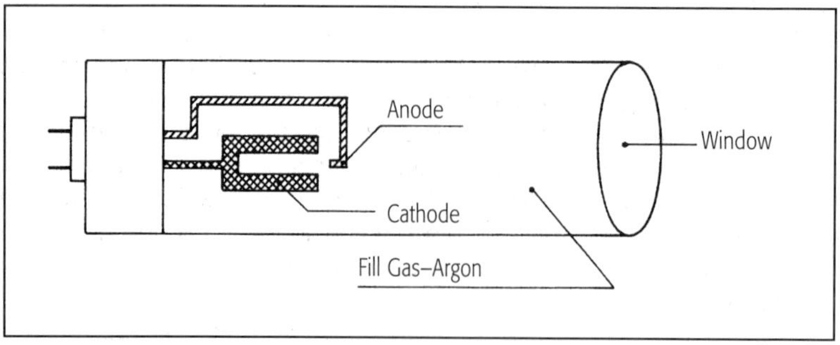

Fig. 1. Hollow cathode lamp.

Table I. **Mechanisms of aluminium atom formation in the furnace.**

graphite ↓

a) $$\underset{(s)}{Me_xO_y} + \underset{(s)}{yC} \longrightarrow \underset{(g)}{xMe} + \underset{(g)}{yCO}$$

b) $$\underset{(s)}{Al_2O_3} \longrightarrow \underset{(g)}{Al_2O_3} \xrightarrow{(1)} \underset{(g)}{Al} + \underset{(g)}{AlO} + \underset{(g)}{O_2}$$

$$\underset{(g)}{AlO} \xrightarrow{(2)} \underset{(g)}{Al} + \underset{(g)}{1/2\ O_2}$$

(1) 234 Kcal mol^{-1}
(2) 114 Kcal mol^{-1}

c) $$\underset{(s)}{Al(NO_3)_3 \cdot 9H_2O} \xrightarrow[(1020\,-\,1270\ K)]{} \underset{(s)}{Al_2O_3}$$

d) $$\underset{(s)}{Al_2O_3 \cdot 6H_2O} \xrightarrow[(770\ K)]{} \underset{(s)}{Al_2O_3}$$

e) $$\underset{(s)}{Al_2O_3} + \underset{(s)}{3C} \longrightarrow \underset{(g)}{2Al} + \underset{(g)}{3CO}$$

f) $$\underset{(s)}{Al_2O_3} + \underset{(s)}{5C} \longrightarrow \underset{(g)}{Al_2C_2} + \underset{(g)}{3CO}$$

(s) solid phase (g) gas phase

have classified aluminium as an element whose atoms become zerovalent by a mechanism which is different from the one suggested for a large number of analyte species (a). Two sequential atomization energies have been obtained for aluminium: 234 Kcal/mol and 114 Kcal/mol (b). Below the temperature at which the atoms appear, Al_2O_3 is a common intermediate for nitrate and chloride (c) (d). Since the reduction of this oxide is not thermodynamically favourable at the appearance temperature, the element must necessarily be formed by the thermic decomposition of the oxide. Other patterns of reaction have also been proposed for aluminium (e) (f). In this case, if the decomposition temperature of the oxide or carbide are not reached before the volatilization of the species occurs, a heavy atom loss can reduce the sensitivity of the process. However, with complex matrixes, all the schemes proposed represent only a pure and simple exemplification of processes which can only difficulty be defined. In fact, the main restrictions in aluminium determination are imposed by the sample matrix or rather, by the size and the nature of the interfering substances which exert an influence on the various analytical parameters.

The most frequent type of interference can be divided into two main classes, i.e. spectral or chemical-physical. Spectral interference occurs when the heating of the matrix in the furnace produces volatilization of substances which cause an undesired absorption of the radiation overlapping the analytical signal. This absorption, called «nonspecific», is of molecular origin and is either produced by the action of undecomposed molecules or by a phenomenon called «light scattering» which is stronger at low wavelengths. Thus, the analytical signal could turn out to be masked or increased by the matrix. Interference of the chemical-physical type is also caused by the matrix which interferes with atomization mechanisms, namely the equilibrium and the kinetics of the formation of atomic vapour. This may occur in two different ways: the first, through the formation of compounds which are more volatile or more refractory with the analyte; the second, by physical interference between the element and the matrix. This interference can be so great as to render it impossible to make a direct calibration against aqueous standards.

Numerous chemical modifications can occur of the vaporized species in the graphite furnace; therefore, the substances that remain in the tube after the charring step may interfere with the production of aluminium atoms. Residual compounds containing chlorine and sulphur can produce the above-mentioned type of interference which has effects on the processes of volatilization and dissociation of aluminium in the vapour phase. In particular, chlorine forms volatile aluminium chloride which, because of this high dissociation energy, may reduce the atomic population at the ground-state. Even changes occuring in the graphite structure after repeated thermic cycles performed at high temperatures can interfere with the atomization process. This last effect may be avoided or significantly reduced by covering the graphite tube with a relatively thin coating of pyrolytic graphite. Refractory elements such as aluminium, which form compounds with very high thermic stability, usually benefit the most from pyrolytic graphite. Moreover, the utilization of suitable chemical matrix modifiers enables the formation of aluminium compounds that are quickly dissociated in the atomization stage. Consequently, the interference problem can be satisfactorily overcome by intervening in the volatilization kinetics. This is possible by modifying the chemical or the physical form, either of the analyte or the matrix, in order to obtain selective volatilizations during thermic treatment in the furnace. This procedure, together with the use of the background correction systems (deuterium or tungsten continuous sources, Zeeman effect) has now produced an analytical compromise which is more than acceptable.

In conclusion, several problems may be encountered in the electrothermal atomic absorption determination of aluminium in biological samples that contain a variety of interfering compounds. For instance, when it is necessary to experiment an analytical procedure for the determination of aluminium in human serum or plasma, some fundamental steps must be considered. These include the sample treatment, the furnace thermic programme and the standardization procedures.

Sample Treatment

Before analysis, several procedures can be used for sample preparation:

a) aluminium extraction by organic solvents after a complexing process;
b) heat treatment with mineral acids in order to completely destroy the organic matrix;
c) direct dilution of the sample with diluents acting as matrix modifiers.

Both the first and the second methods are generally not suitable because they can introduce numerous contamination factors which are difficult to control. Moreover, they make the analysis take a much longer time. In the third method, which is now the one most widely used, the operative procedure can vary depending on whether the l'Vov platform at stabilized temperature is utilized or not. In this case (Fig. 2), the sample is deposited on a pyrolytic platform which is inserted in the graphite tube and heated exclusively by radiation from the tube wall. Consequently, the platform is heated at a slower rate than the tube, the atomic vapour of the aluminium is produced at a relatively high and constant

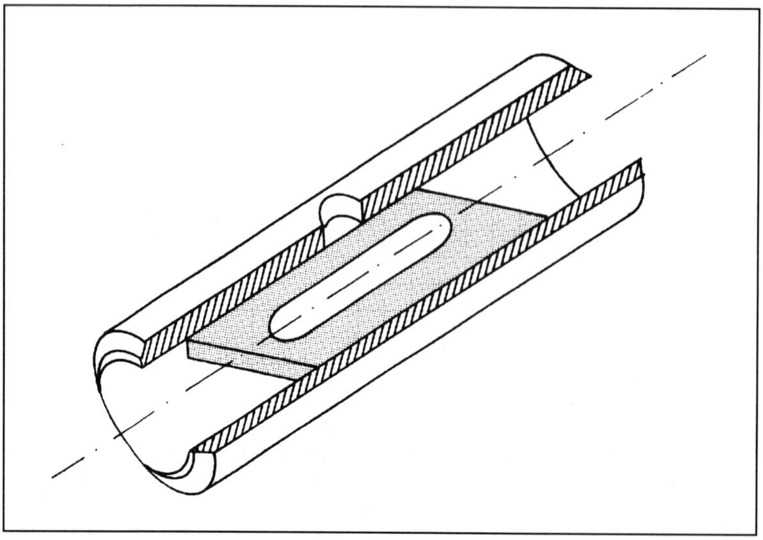

Fig. 2. Graphite tube with l'Vov platform inserted.

temperature, and the analytical peak is actually proportional to the number of the atoms.

In the furnace, a series of problems can also be encountered:

- the sample drying may not be uniform;
- carbon build-up can form on the platform after a certain number of determinations and interfere with the ones that follow.

Often, the general result of these drawbacks is a loss in sensitivity and reproducibility. However, these problems can be controlled with a suitable thermic programme. This represents the second important parameter of the whole analytical procedure.

Furnace Thermic Programme

An increase of temperature which is too rapid during the drying step is the major cause of an irregular process which can also result in the sample sputtering inside the tube. The charring step must completely remove or destroy the matrix by way of a combination of times and temperatures, and at the same time avoid carbon build-up on the platform. Since temperatures and times represent a critical point in the thermic programme, it is advisable to carry out a study of the thermic profile in order to select the best operative conditions for the charring step (Fig. 3). The same procedure should be performed in order to find the correct atomization temperature; however, this parameter is less critical than the previous one. Aluminium atomization normally occurs at the temperatures which are higher than those required by other metals. This is because the element needs more energy in order to produce the required atomic population.

Standardization Procedures

As far as the calibration method is concerned, some differences of opinion exist between authors who assert that the calibration should be carried out against aqueous standards[14,15] and others authors who have demonstrated the need to use the standard addition method.[12,16] In the course of our experience, we have sometimes encountered difficulty with the direct calibration method carried out in absence of the serum matrix. Moreover, even the slopes of standard curves constructed by using sera from normal people and sera from uremic patients resulted in being very different. Therefore, it appears that it is best to use the standard addition method, even if it is also possible to calibrate against curves constructed in the presence of a matrix which is similar to the one present in the sample. However, it must be kept in mind that

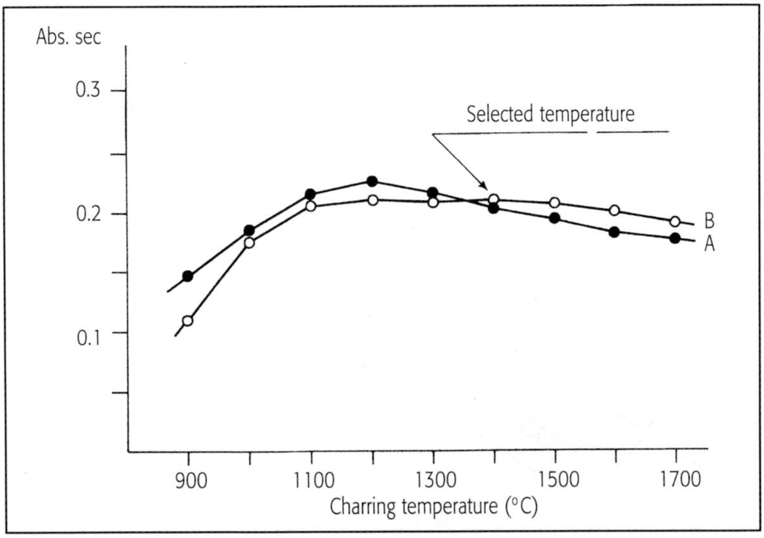

Fig. 3. Effect of the charring temperature on the aluminium absorption signal.

Serum sample diluted 1+1

Diluent:
A = H_2O
B = $Mg(NO_3)_2$ 2.5 g/L + Triton-X 100 0.05%

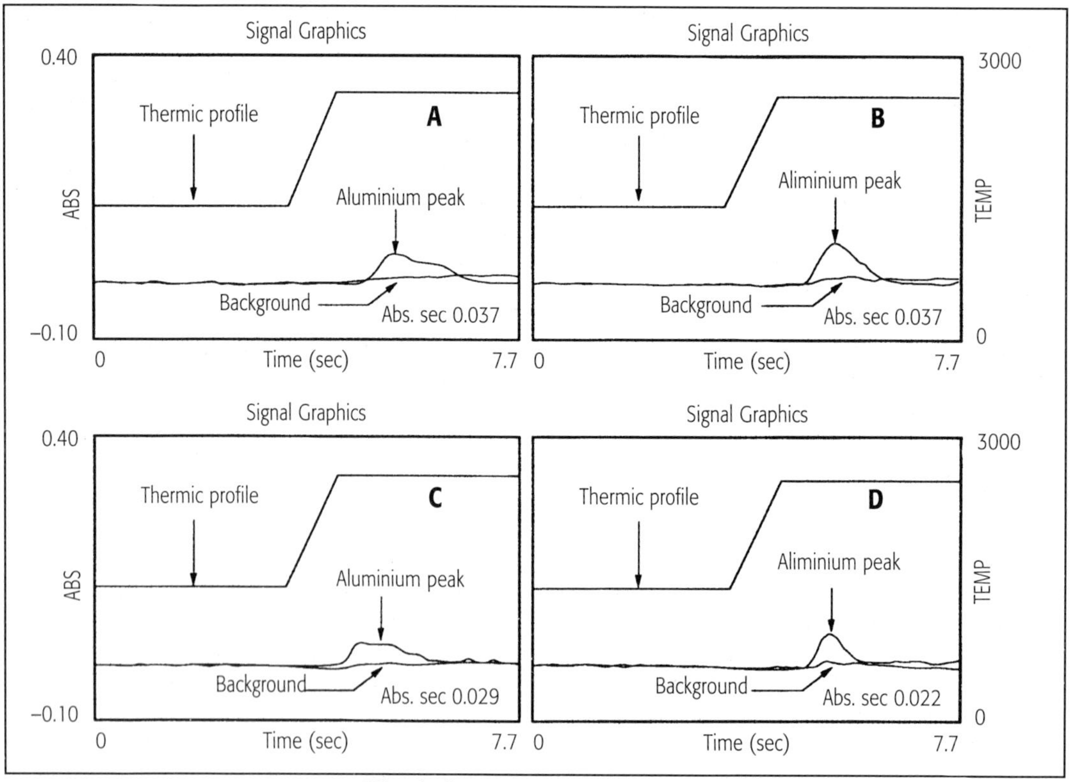

Fig. 4. Serum aluminium peaks.

Aluminium concentration in serum: 40 μg/L
Dilution ratio: 1:4
Matrix modifier:
A – $Mg(NO_3)_2$ 2.5 g/L
B – $Mg(NO_3)_2$ 2.5 g/L + Triton-X 0.05%
C – HNO_3 1% v/v + Triton-X 0.1%
D – $NH_4H_2PO_4$ 5 g/L + Triton-X 0.05%

the analytical response can also vary according to the dilution ratio of the sample. For these aspects, when recordering the analytical peak, the use of peak height results in being very restrictive. The integrated peak area, together with the utilization of a stabilized temperature platform, makes it possible to considerably reduce the risks of error which are correlated to the variation of the peak shapes that are characteristic of a progressive wear and tear of the graphite. The standard addition method also requires that the linearity range of analytical response be checked. This factor, which usually dependents both on the nature of the element and its chemical matrix, is sometimes influenced by instrumental characteristics. This mainly occurs when old spectrophotometers are coupled with the latest models of furnaces which produce very fast signals in the atomization stage. This results in a progressive loss of linearity as the concentration of the elementincreases.

In conclusion, operative conditions should be experimentally chosen both on the basis of the instrumental characteristics and the procedure adopted. The following is a general scheme for aluminium determination in serum:

a) Analytical wavelength: 309.3 nm (a secondary wavelength at 396.2 nm extends the linearity range of aluminium concentration

Table II. **Matrix modifiers for ETA-AAS aluminium determination.**

$Mg(NO_3)_2$ + Triton-X 100
H_2SO_4 + Triton-X 100
HNO_3 + Triton-X 100
EDTA + NH_4OH + H_2SO_4
$NH_4H_2PO_4$ + Triton-X 100
$K_2Cr_2O_7$
NH_4NO_3 + Triton-X 100

but reduces the sensitivity by about 25%).

b) Matrix modification by diluting the serum sample with a solution of magnesium nitrate 2.5 g/L which slows aluminium vaporization and makes it possible to use a higher charring temperature. Adding a detergent, such as Triton-X (0.05%), improves the precision of sample handling. The dilution ratio is a function of the aluminium concentration in the sample. For this reason, the dilution must be carried out so that the concentration remains within the linearity range. Alternative matrix modifiers are reported in Table II. Aluminium peaks obtained my means of some matrix modifiers are shown in figure 4.

c) Use of the l'Vov platform and the integrated peak area.

d) Use of background correction systems.

e) Calibration carried out by the standard addition method or in presence of serum matrix.

A suitable furnace thermic programme is shown in Table III. The results of data quality tests obtained by using the parameters described above are reported in Table IV.

In general, the considerations so far expressed are valid for the determination of aluminium in all biological matrixes. However, it is evident that inorganic matrixes can require changes both in the thermic programme and/or the matrix modification parameters.

For example, when determining the element in dialysis fluids, which have a particularly concentrated inorganic matrix (Table V), a good matrix modification can be obtained by diluting (1 : 5) the sample with nitric acid 1% v/v + Triton-X 100 0.1%. A thermic programme, optimized experimentally by using the previously described procedures, is reported in Table VI.

In order to elucidate the biological role of the element,[17] it may be necessary to determine aluminium in tissues, particularly bone tissue.[18] Unlike serum, for which numerous methods have been proposed in the last few years, only a restricted number of studies have been carried out to measure aluminium concentration in these matrixes.[19-21] In this case, the analysis is complicated by the required sample treatment, unnegligible interference and, sometimes, by sample exiguity (e.g. human bone biopsy). Acid digestion is the most widely used method for the distruction of the organic matrix prior to instrumental determination; however, in order to reduce contamination from the added acids, a dry ashing procedure is preferable. Since a more compli-

Table III. **Furnace programme for aluminium determination in serum.**

Step	Temperature (°C)	Ramp time (sec)	Hold time (sec)
I Drying	90	30	10
II Drying	130	10	10
I Ashing	700	20	10
II Ashing	1400	20	10
Atomizing	2500	0	4
Cleaning	2650	1	5
Cooling	20	1	20

Gas flow: argon (stopped in the atomization stage). Sample volume: 10 μL.

Table IV. **Data quality tests.**

Detection Limit: 0.5 μg/L (3 × SD as calculated on the basis of 10 measurements of the reagent blank).
Quantification Limit: 1.6 μg/L (10 × SD as calculated on the basis of 10 measurements of the reagent blank).
Sensitivity: 1.2 μg/L (concentration giving 0.0044 absorbance units).
Accuracy: 97 – 104% (recovery performed on reference material).
Precision: Al concentration 5 μg/L: Coeff. of variation 3.8% Al concentration 50 μg/L: Coeff. of variation 2.4%

Table V. **Typical dialysis fluid composition.**

Type A		**Type B**	
Sodium Chloride	g 214.63	Sodium Chloride	g 208.66
Potassium Chloride	g 4.03	Potassium Chloride	g 5.22
Magnesium Chloride	g 5.48	Magnesium Chloride	g 5.34
Sodium Acetate	g 186.00	Sodium Acetate	g 181.00
Calcium Chloride	g 10.60	Calcium Chloride	g 10.29
		Glucose	g 35.00

Table VI. **Furnace programme for aluminium determination in dialysis fluids.**

Step	**Temperature (°C)**	**Ramp time (sec)**	**Hold time (sec)**
I Drying	90	10	10
II Drying	130	20	10
I Ashing	500	40	20
II Ashing	800	10	10
III Ashing	1400	20	20
Atomizing	2500	0	5
Cleaning	2650	1	5
Cooling	20	1	20

Gas flow: argon (stopped in the atomization stage). Sample volume: 10 μL.

cated sample manipulation is required than in the ones of serum or dialysis fluids, particular attention must be taken regarding contamination. Sample solutions can be analyzed either by ICP or atomic absorption spectrometry by electrothermal atomization.[22] Using the last method, the matrix can suitably be modified by a solution of 5 g/L magnesium nitrate. The sample/diluent ratio must be chosen in relation to the aluminium content.

Contamination

Aluminium is an ubiquitous element and its presence in ordinary dust is well known. For this reason, contamination represents a serious problem in the determination of the element. Generally, it can be divided into two potential sources of external aluminium: migration from the glass and environmental contamination during sampling and analysis procedures. Therefore, several precautions must be adopted. The use of glassware should be minimized and the whole equipment used for specimens should be checked before introduction into the routine. Two different aspects must be considered:

1) ***Sample Contamination Outside of the Analysis Laboratory***

Usually this problem is linked to sampling and sample storage: syringes, needles, test-tubes.

2) ***Sample Contamination Inside the Analysis Laboratory***

A series of factors may be responsable for this:
- Test-tubes used in the sample handling.
- Autosampler cups.
- Autosampler tip.
- Pipette tips.
- Use of glass material, especially with acid solutions.
- Distilled water used in sample dilution.
- Chemical reagents.
- Laboratory cleanliness: dust, smoke, cleaning of air conditioner filters.

Reproducibility of the Measurements

Another series of factors is responsible for reproducibility problems, such as:
- Bad quality of the sample.
- Incorrect sample deposition on the platform.
- Incorrect position of the autosampler tip.
- Carbon build-up.

References

1. Alfrey A.C., Hegg S., Craswell P.: *Am. J. Clin. Nutr.*, **33**, 1509-1516, 1980.
2. Cournot-Witmer G., Zingraff J., Bourdon R., Drüeke T., Balsan S.: *The Lancet*, **ii**, 795-796, 1979.
3. Drüeke T.: *J. Nephrol.*, **II**, 45-58, 1989.
4. Crapper D.R., Krishnan S.S., Quittkat S.: *Brain*, **99**, 67-80, 1976.
5. Candy J.M., Oakley A.E., Klinowski J., Carpenter T.A., Perry R.H., Atak E.J.R., Perry E.K., Blessed G., Fairbairn A., Edwardson J.A.: *The Lancet*, **i**. 354- 357, 1986.
6. Martin R.B.: in Siegel H., Siegel A. eds. *Metal Ions in Biological Systems*, vol. 24, Dekker M., Pubbl. New York and Basel, 1988, p. 2-57.
7. Ganrot P.O.: *Environ. Health Perspectives*, **65**, 363-441, 1986.
8. Hem J.D.: *Kidney Int.*, **29**, suppl. 18, S3-S7, 1986.
9. Cawse P.A.: UK Atomic Energy Authority Report AERE R-9484, HMSO, London, 1978.
10. Sherlock J.C.: in Massey R.C., Taylor D. eds.: *Aluminium in Food and the Environment*, **73**, 68-76, 1988. The Royal Society of Chemistry, Thomas Graham House, Publ., Cambridge.
11. Bellomonte G.: *Ped. Oggi*, vol. IX, **6**, 131-132, 1989.
12. Savory J., WIlls M.: *Kidney Int.*, **29**, suppl. 18, S24-S27, 1986.
13. Sturgeon R.E., Chakrabarti C.L., Langford C.H.: *Anal. Chem.*, vol. 48, **12**, 1793-1807, 1976.
14. Xiao-Quans S., Shen L., Zhe-Ming N.: *J. Anal. At. Spectr.*, **3**, 99-103, 1988.
15. Bettinelli M., Baroni U., Fontana F., Poisetti P.: in Minoia C., Caroli S., eds. *Applicazioni dell'ETA-AAS Zeeman nel laboratorio chimico e tossicologico*, vol. 2, Libreria Cortina Publ., Padova, 467-481, 1989.
16. Costantini S., Giordano R., Vernillo I.: *Microch. J.*, **30**, 425-434, 1984.
17. Costantini S., Giordano R., Ippolo A., Mantovani A., Ballanti P., Mocetti P., Bonucci E.: *Pharmacology and Toxicology*, **64**, 47-50, 1989.
18. Ballanti P., Mocetti P., Della Rocca C., Costantini S., Giordano R., Ioppolo A., Mantovani A., Stasolla D., Bonucci E.: *Min. Metab. Res. in Italy*, **4**, 139-142, 1983.
19. Boyce B.F., Elder H.Y., Fell G.S., Nicholson W.A.P., Smith G.D., Dempster D.W., Gray C.C.,Boyle I.T.: *Scanning Electron. Microsc.*, **3**, 329-337, 1981.
20. Goode G.C., Herrington J., Goddard P.C.: *Radiochem. Radioanal. Lett.*, **31**, 87-94, 1977.
21. LeGendre G.R., Alfrey A.C.: *Clin. Chem.*, **22**, 53-56, 1976.
22. Giordano R., Costantini S., Vernillo I., Casetta B.,, Aldrighetti F.: *Microch. J.*, **30**, 435- 447, 1984.

Section II

BIOLOGY

Aluminum and Membrane Channels

M. Colombini

Introduction

Cell membranes contain a variety of transport systems which control the flow of ions and small molecules from one cellular compartment to another. One such system is the channel-forming protein. This is a protein that is embedded in the membrane and forms a continuous water-filled pathway through the membrane. Such a pathway would tend to dissipate electrochemical gradients and thus is generally under some sort of control.

Gating refers to the variety of processes that exist in different membrane channels to control their permeability. Some are gated by the membrane potential (voltage gated), some by small molecules (chemical gated), and some by the tension in the membrane (stretch gated). Gating relies on the channel-forming protein existing in different conformational states differing in their ability to allow molecules and/or ions to cross the membrane. The gating process allows relatively small changes in the environment to cause dramatic shifts in the state occupied by a given channel. For example, in the case of a voltage-gated channel, the channel may have a high probability of being in an open or highly-permeable state in the absence of a membrane potential and a low probability of being in the open state in the presence of even a small potential. For these gating processes to be useful physiologically, the different conformational states must be very close in energy, often differing only by a fraction of the energy of a hydrogen bond.

Metal ions such as aluminum could, in principle, interfere with finely-tuned systems such as gated membrane channels if they bind to these proteins. This is reasonable from hindsight, but when effects of aluminum on voltage-gated channels were first observed, they were a big surprise. Indeed, they were discovered by accident.

Effects of Aluminum Salts on Membrane Channels

Mitochondria from all eukaryotic kingdoms[1] contain channels in their outer membranes that allow ions and small molecules to diffuse between the cytoplasm and the mitochondrial spaces. These channels are called VDAC, an acronym for voltage-dependent anion-selective channel. As the name indicates, these channels respond to the membrane potential and enter closed conformations when a voltage is applied. This voltage- gating may allow these channels to control mitochondrial function by controlling the permeability of the outer membrane.

VDAC channels have a relatively simple structure. They form 3 nm diameter pores with just one 30 kDa polypeptide.[2] In the outer membrane of *N. crassa* they can be induced[3] to form two-dimensional crystals (Fig. 1) whose structure consists of a 6-channel repeating unit. At the resolution of the images, the surface structure of the crystal seems the same from both sides reflecting the symmetrical electrophysiological behavior.

Less than 10 μM $AlCl_3$ causes a reduction in the voltage dependence[4] of VDAC.[1] At higher levels one observes a profound change in the ability of VDAC to respond to a membrane potential. These channels are normally open at low membrane potential and close when positive or negative voltages are applied (the switching region is around $\pm$ 20 mV).

Channel closure can be observed as a decay in ion flow through the channels as a function of time after the application of a membrane potential (Fig. 2). VDAC channels were reconstitued into planar phospholipid membranes separating two aqueous compartments. This allows the properties of the channels to be studied without the influence of other cellular components. The composition of the solution

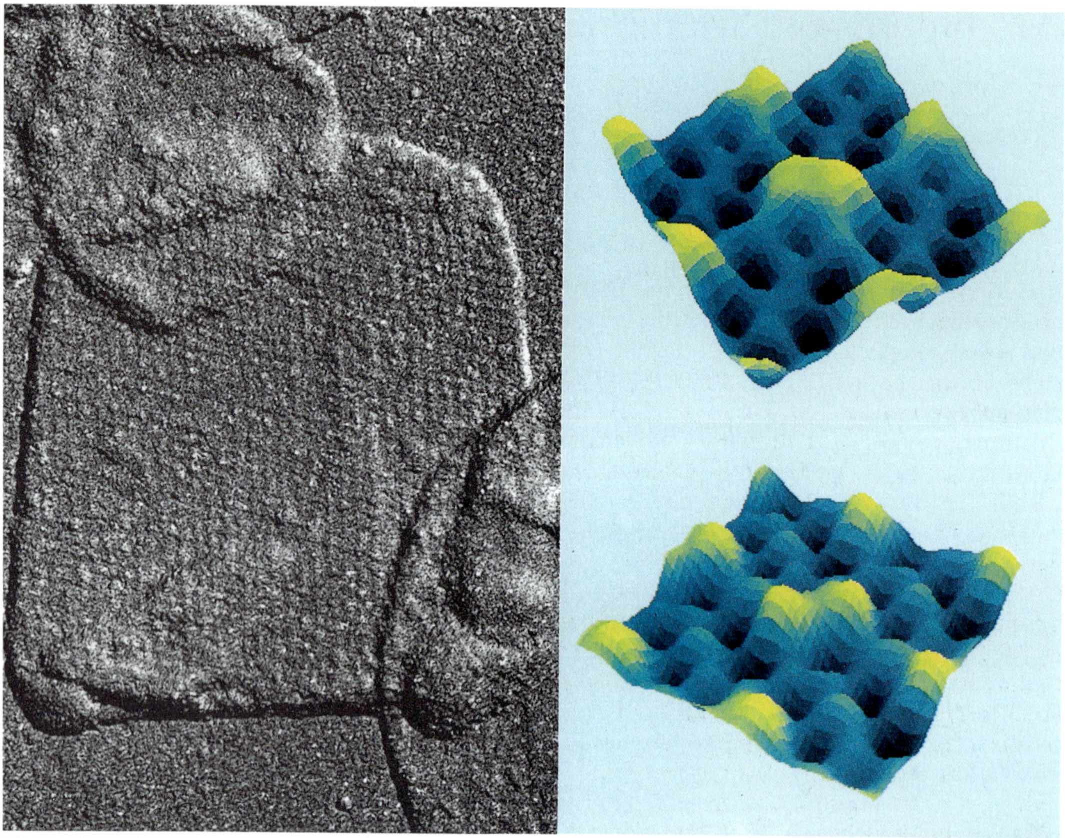

Fig. 1. Two-dimensional crystals of VDAC channels in outer mitochondrial membranes of *N. crassa*. The left panel is an electron micrograph of freeze-dried and shadowed outer membranes after treatment with phospholipase A_2 to induce large crystalline arrays. The crystalline pattern is visible on the surface of the central flattened vesicle with straight edges. On the right are the results of computer filtration, averaging and image reconstruction. The false color indicates elevation (bright yellow being high and dark blue being low). The two images represent the two surfaces of the crystal. Each image contains four six-channel repeating units. The figure is adapted from the work of Thomas *et al.*[2] The color images was produced by B.L. Trus.

on either side of the membrane can be controlled exactly and so can the membrane potential. The potential can be used to drive ions through the channels and the resulting current can be used to access the permeability of the population of channels in the membrane. If a closure-inducing membrane potential is applied, the current through the population of channels decreases with time as channels undergo conformational changes from the highly-conductive, open, state to a low-conducting, closed, state.

The rate channel closure and the extent of closure were dramatically reduced after the addition of $AlCl_3$. A remarkable aspect of these findings is that the experiments were performed at pH 7 where the concentration of free Al^{3+} is vanishingly small due to its hydration in solution to hydroxylated forms. Even more remarkable was the fact that increasing the free concentration of Al^{3+} by many orders of magnitude by reducing the pH to 4 (keeping the total aluminum constant) essentially eliminated the effect of aluminum (Fig. 3). This indicated that the species that acts on VDAC is not Al^{3+} but some other form of aluminum

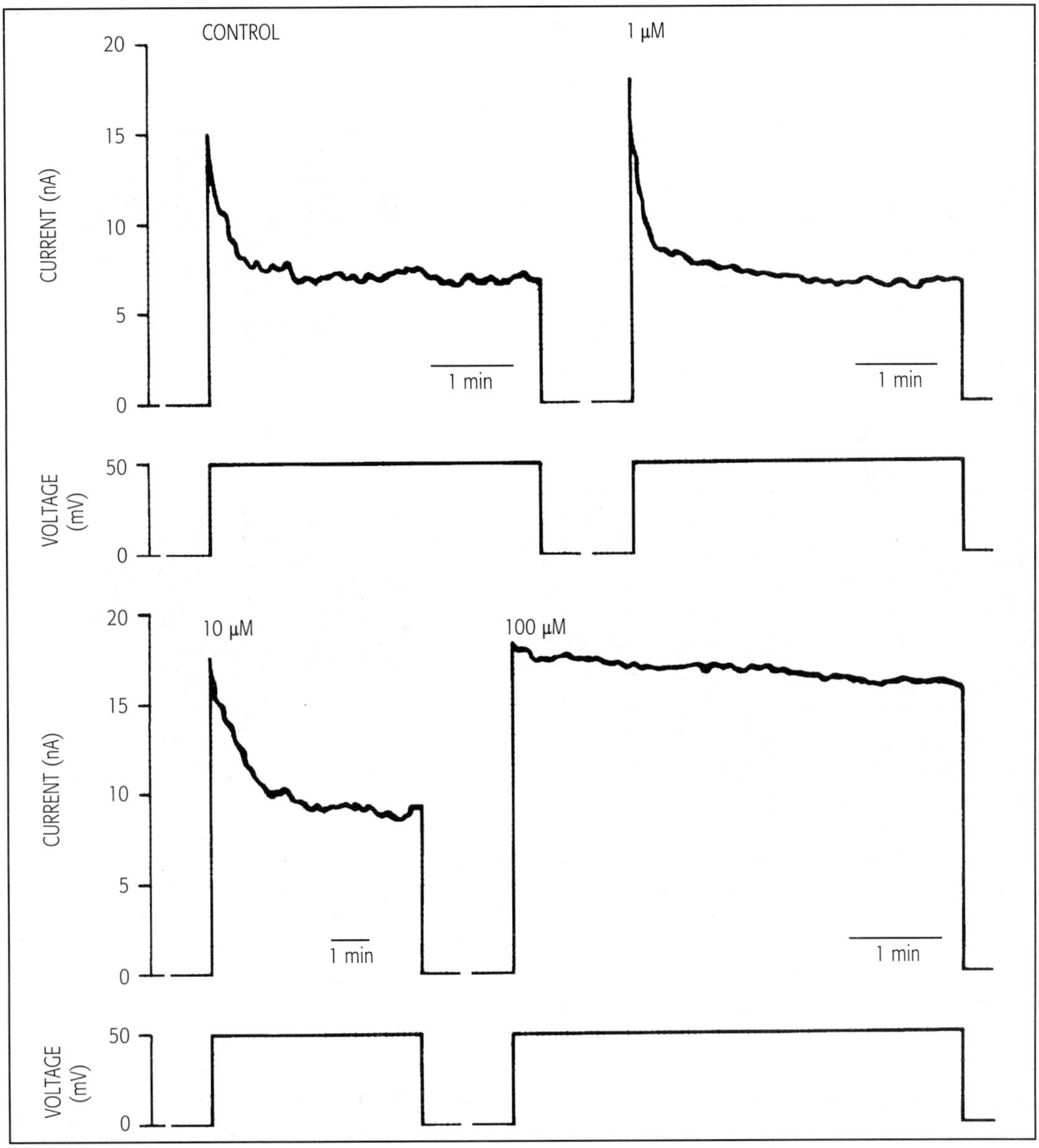

Fig. 2. Low levels of $AlCl_3$ decrease both the rate and extent of voltage-dependent channel closure. Sequential additions of $AlCl_3$ were made to a planar phospholipid membrane (soybean phospholipids) containing many VDAC channels isolated from the mitochondria of *N. crassa*. The current levels were recorded in response to the applied voltage as indicated. Reproduced with permission from Dill *et al.*[4]

more common at physiological pH.

Experiments on neuroblastoma cells[5] show that Pb^{2+}, Cd^{2+}, and Al^{3+} additions open up channels in their surface membranes. These channels were normally closed and induced to open by the addition of these metals. Although the authors speculate that this may be a metal ion activated channel, the results are quite similar to those obtained with VDAC. The ability of the channels to remain closed

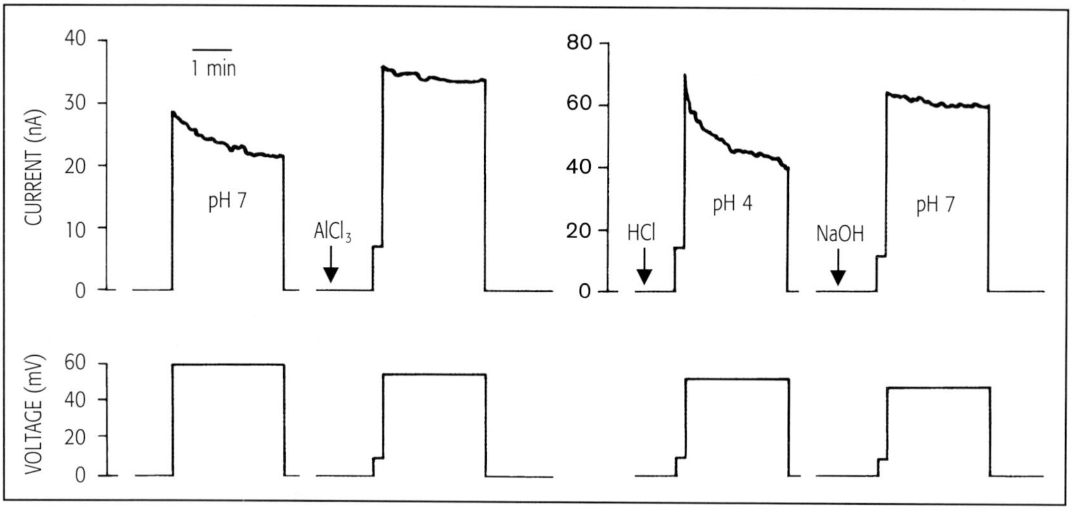

Fig. 3. Aluminum's inhibition of VDAC's voltage dependence is reversed at acid pH. VDAC channels from *N. crassa* mitochondria were inserted into a planar phospholipid membrane made using diphytanoyl phosphatidylcholine. 100 μM $AlCl_3$ was added to the solution on both sides of the membrane (1 M KCl, 5mM $CaCl_2$, 10 mM Tris-HCl, initially pH 7). The records were all obtained in the same experiment. The breaks in the record indicate that only the relevant portions of the record are displayed. Since channels kept on inserting throughout the experiment, the scale had to be changed in the middle. Reproduced with permission from Dill *et al.*[4]

under normal conditions is inhibited by the presence of aluminum. The experiments were performed at pH 7.2 indicating that a species other than Al^{3+} is probably involved but the authors did not explore this possibility.

The ability of aluminum fluoride to activate G proteins which can, in turn, affect membrane channels, has been established and will not be discussed here because the high levels of fluoride needed are not physiological.

Aluminum Hydroxide as an Active Species

When aluminum salts affect cellular processes the species that seems most likely to be responsible is Al^{3+}. Its high field strength allows it to interact strongly with polar molecules. However, this high field strength also results in the hydration of Al^{3+}. At neutral pH, most of the aluminum exists in the form $Al(H_2O)_3(OH)_3$.[6] In addition, mono, di, and tetrahydroxylated species also form and dominate at lower and higher pH levels (Fig. 4). In principle, any of these species could bind to and affect cellular macromolecules.

Upon noticing that at acidic pHs, the aluminum effect on VDAC could be virtually eliminated, it became important to address the question of which species was actually acting on this membrane channel. The effect of pH alone is insufficient because one cannot distinguish between pH effects on the channel and effects on the aluminum species in solution. By examining the ability of other trivalent ions to mimic the action of aluminum one might be able to separate pH effects from effects of altering the level of any given hydrated species.

Gallium and indium are in the same group as aluminum in the periodic table but the levels of their hydroxylated species at a given pH are quite different from that of aluminum (Fig. 4). This is reflected in the amounts of aluminum, gallium or indium which must be added to achieve comparable effects.[7] At pH 5, 40μM of $AlCl_3$ or $InCl_3$ are needed to obtain results comparable to those achieved with just 1μM

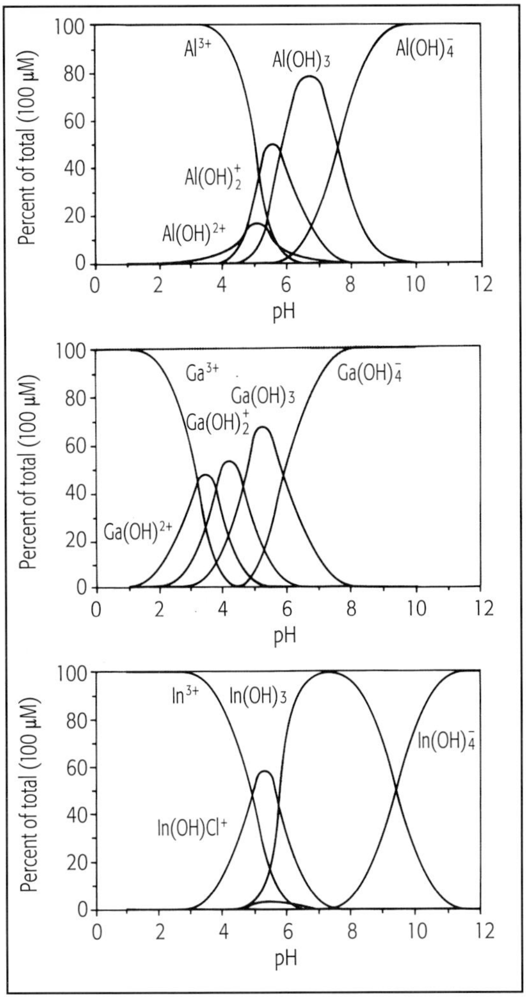

Fig. 4. Distribution of the hydrolysis products of aluminum, gallium, and indium as a function of pH. Formation constants were obtained from Baes & Mesmer[6] and corrected for ionic strength, osmotic coefficients, and molality of all anions as they described. The concentrations of the species at a particular pH were calculated by solving simultaneously equations for all the known complexes present in solution. The small peak in the indium graph is $In(OH)_2^+$. Reproduced with permission from Zhang & Colombini.[7]

$GaCl_3$ (Fig. 5). By contrast, at pH 7, indium and aluminum were 10 times more potent than gallium. When the potency observed in any experiment was plotted vs. the log of the concentration of each metal species in solution, the potency was pH and metal independent if the species chosen was the trihydroxide (Fig. 6). If, the free trivalent metal ion were the active species, depending on conditions, comparable effects could be achieved by free metal ion concentrations differing by 8 orders of magnitude! This seems highly unlikely.

A further indication that the free trivalent metal ion is not the active species comes from the observation that La^{3+}, an ion that does not form significant amounts of hydrated complexes in solution, is completely ineffective even at concentrations two order of magnitude greater than those needed to obtain effects with aluminum (Table I).

Other ions (Sc^{3+}, Cr^{3+}, and Fe^{3+}) capable of generating trihydroxide complexes were also able to mimic aluminum's effect on VDAC. The minimum amounts needed to observe effects varied from metal ion to metal ion but, remarkably, correlated very well with regard to the amount of trihydroxide species formed in solution (Table I).

These results strongly support the conclusion that aluminum trihydroxide is the active species. This is surprising not only because it is a neutral species but also because it is very poorly soluble. However, solubility is measured at equilibrium and systems are often away from equilibrium. Most cellular processes are out of equilibrium. The addition of small amounts of an acidic $AlCl_3$ solution to a soluzion buffered at neutrality did not result in a visible precipitate as expected at equilibrium. No precipitate was detected days later indicating a stable, supersatured solution. Such was not the case with $CrCl_3$ which precipitated in a short time resulting in loss inhibition of VDAC closure.[7]

Multiple Aluminum Hydroxide Binding Sites on VDAC

If aluminum chloride was added to both sides of a membrane containing VDAC channels, the channels became much less sensitive to the membrane potential. The channels tended to remain in the open state. However, if the aluminum salt were added to just one

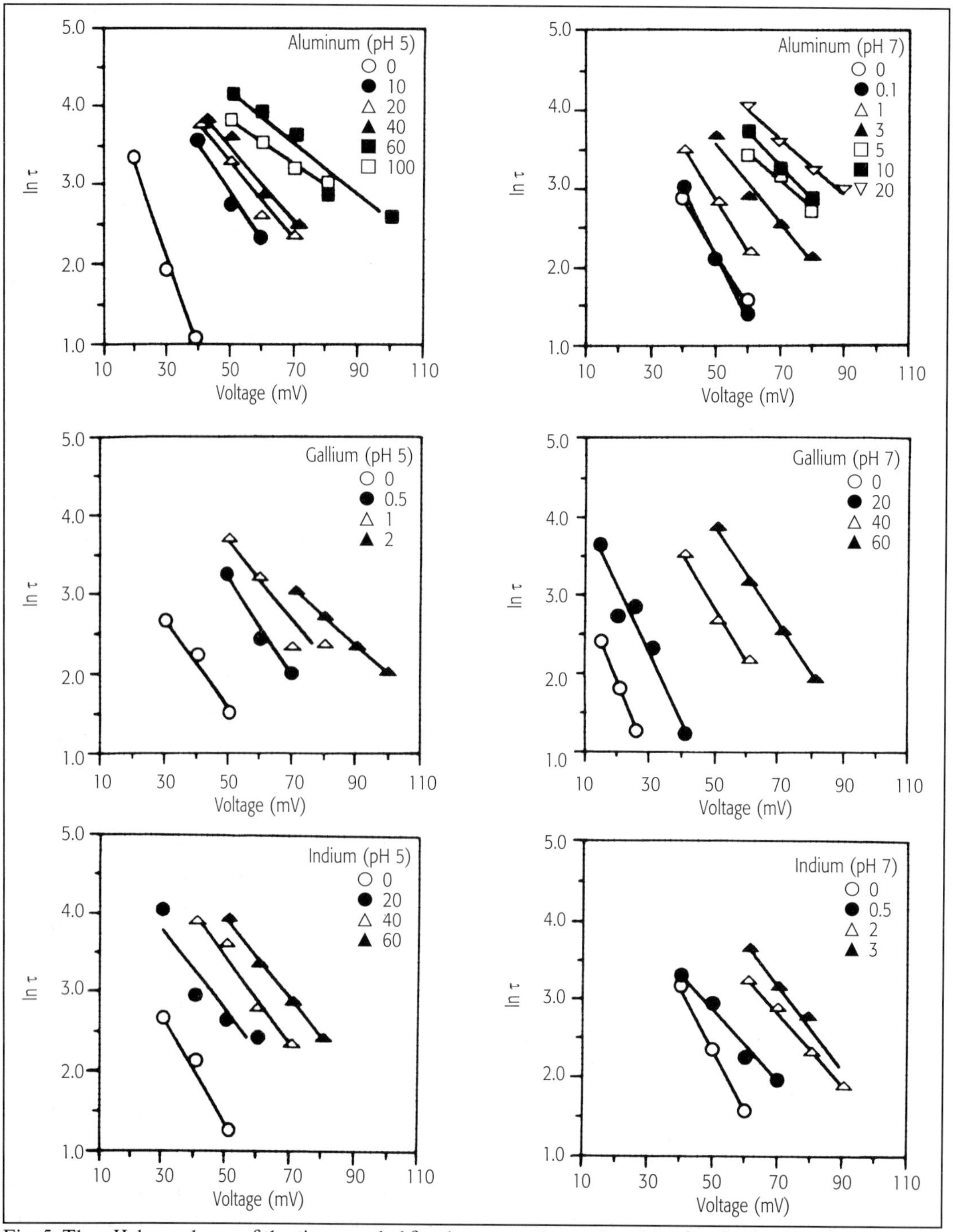

Fig. 5. The pH dependence of the time needed for the current to reach to 1/e of its final value after the application of the indicated potential (for similar records as those shown in Fig. 2). Each panel shows a separate experiment performed under the indicated conditions on a single channel- containing membrane. Sequential additions (to both sides of the membrane) of metal salts were made to final bath concentrations as indicated (in μM). Reproduced with permission from Zhang & Colombini.[7]

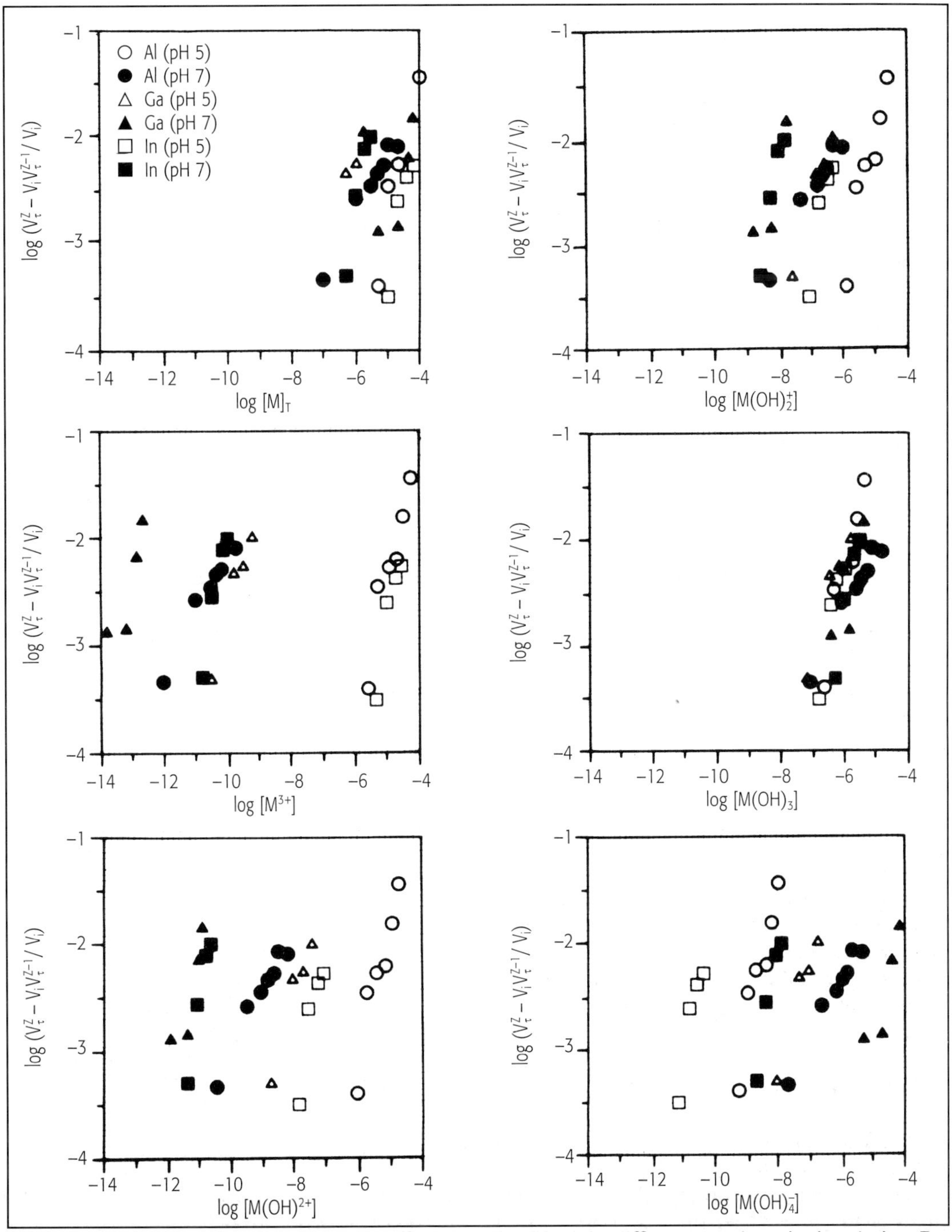

Fig. 6. Correlations between the inhibitory effects on VDAC and the different metal species in solution. Results such as those shown in Fig. 5 were used to construct these panels. The ordinate is the result of a transformation of the data according to a theory[7] in order to obtain a parameter that represents the effect of aluminum but should vary linearly with the log of the concentration of the species. M_T is the total metal salt added. Reproduced with permission from Zhang & Colombini.[7]

Table I. **Comparison of the characteristics of effective and ineffective metals.**

Ion (M^{z+})	Surface Charge Density (e/A^2)	Ionic Radius (Å)	Lowest Effective or Highest Tested Conc.	
			$M(OH)_z$* (M)	Total (M)
Effective				
$_{13}Al^{3+}$	0.92	0.51	8×10^{-7}	1×10^{-6}
$_{21}Sc^{3+}$	0.45	0.73	7×10^{-7}	1×10^{-5}
$_{31}Ga^{3+}$	0.62	0.62	4×10^{-7}	5×10^{-6}
$_{49}In^{3+}$	0.36	0.81	5×10^{-7}	5×10^{-7}
$_{24}Cr^{3+}$	0.60	0.63	1×10^{-7}	5×10^{-4}
$_{26}Fe^{3+}$	0.58	0.64	$<4\times10^{-4}$	5×10^{-4}
Ineffective				
$_{12}Mg^{2+}$	0.37	0.66	--	5×10^{-3}
$_{20}Ca^{2+}$	0.16	0.99	--	5×10^{-3}
$_{27}Co^{2+}$	0.31	0.72	3×10^{-9}	5×10^{-4}
$_{29}Cu^{2+}$	0.31	0.72	$<6\times10^{-8}$	5×10^{-4}
$_{30}Zn^{2+}$	0.29	0.74	3×10^{-6}	5×10^{-3}
$_{57}La^{3+}$	0.23	1.02	0.✦	5×10^{-4}

* Calculated according to Baes and Mesmer[6] from lowest effective concentration for effective metals and highest concentration tested for ineffective metals at pH 7 (pH 6.6 for chromium), ionic strength = 1.

✦ Essentially not formed at pH 7.

side, the channels would close poorly if the aluminum side were made negative but close very well if the aluminum side were made positive. Indeed, positive potentials applied to the aluminum-containing side resulted in channel closure and poor channel reopening.[8] This result is illustrated (Fig. 7) with indium as the effector.

This membrane contained a few channels so that individual channel closures were easily visible. The channels closed with positive and negative membrane potentials prior to indium addition. In the presence of 2 μM $InCl_3$, channels fail to close when the indium-free side was made positive (i.e. negative on the indium-containing side). However, with a negative potential closure was both rapid and extensive. A return to −10 mV, a potential which allowed the channels to reopen prior to indium addition, resulted in a lower current (compare levels at two adjacent arrows) indicating that most of the channels remained closed. The reapplicatrion of a positive potential also resulted in a smaller current because some channels were still closed. Positive potentials usually induce the channels to reopen if these are sufficiently large and applied for long-enough time.

The asymmetric behaviour resulting from asymmetric aluminum addition can be understood in terms of two aluminum binding sites able to translocate through the membrane (Fig. 8). Aluminum added to one side binds to a site that would move to the other side if a negative potential were applied to the aluminum-containing side. The binding of aluminum prevents this and so the channels do not

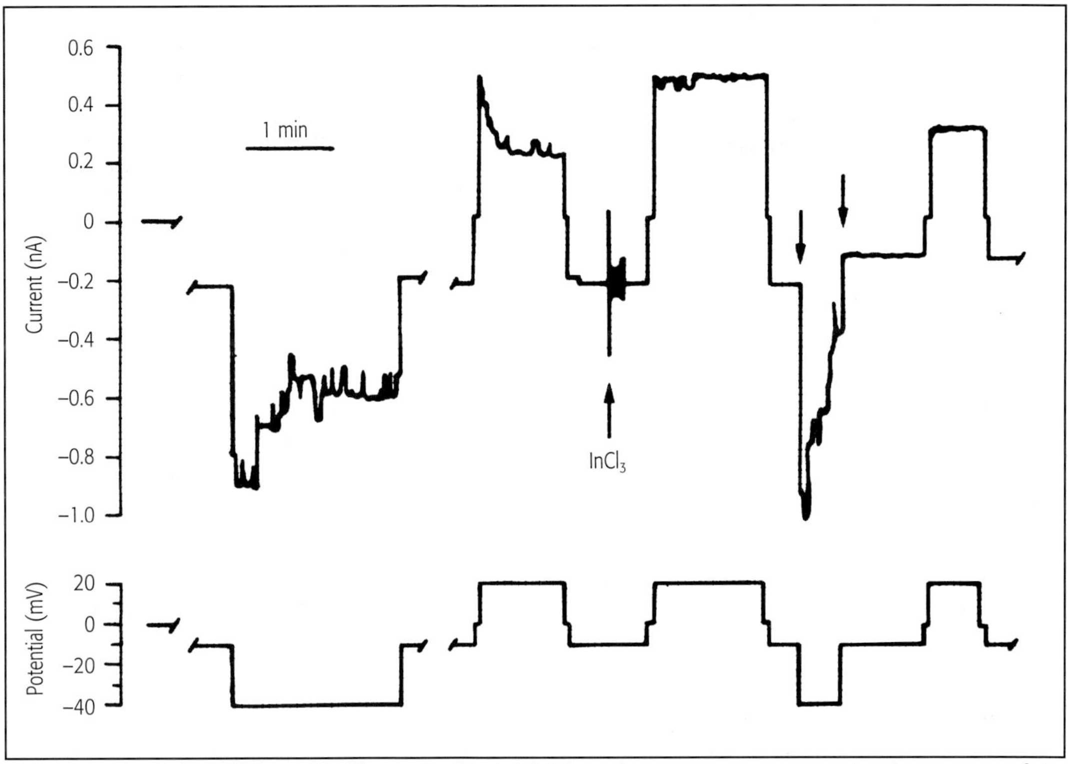

Fig. 7. Asymmetric addition of $InCl_3$ resulted in inhibition of one gating process and enhancement of the other. Experimental conditions were similar to those in Fig. 2. $InCl_3$ was added to one side of the membranes as indicated to a final conc. of 2 μM. The sign of the applied potential refers to the side of the membrane lacking indium. Reproduced with permission from Zhang & Colombini.[8]

close. However, the channels do close if a positive potential is applied because another domain is moving from the opposite side to the aluminum-containing side. Once there, aluminum binds inhibiting the re-opening of the channel. While other explanations may account for the observations, this is the most straight-forward interpretation.

Conclusions

Although investigations on the effects of aluminum salts on membrane channels are at an embryonic stage, it is already clear that the effects observed to date are among the most potent effects of aluminum so far described. Aluminum opens the channels or keeps them open presumably by inhibiting the gating processes that regulate these channels. At least in the case of VDAC channels from mitochondria, the effects are mediated by a neutral species (aluminum trihydroxide) most prevalent at physiological pH. Only time will tell how widespread the phenomenon is and whether aluminum hydroxide turns out to be the active species in other aluminum-influenced cellular processes.

Acknowledgements. I am grateful to L. Thomas and B.L. Trus for contributing Figure 1. This work was supported by the Office of Naval Research.

References

1. Colombini M.: *J. Membrane Biol.*, **111**, 103-111, 1989.

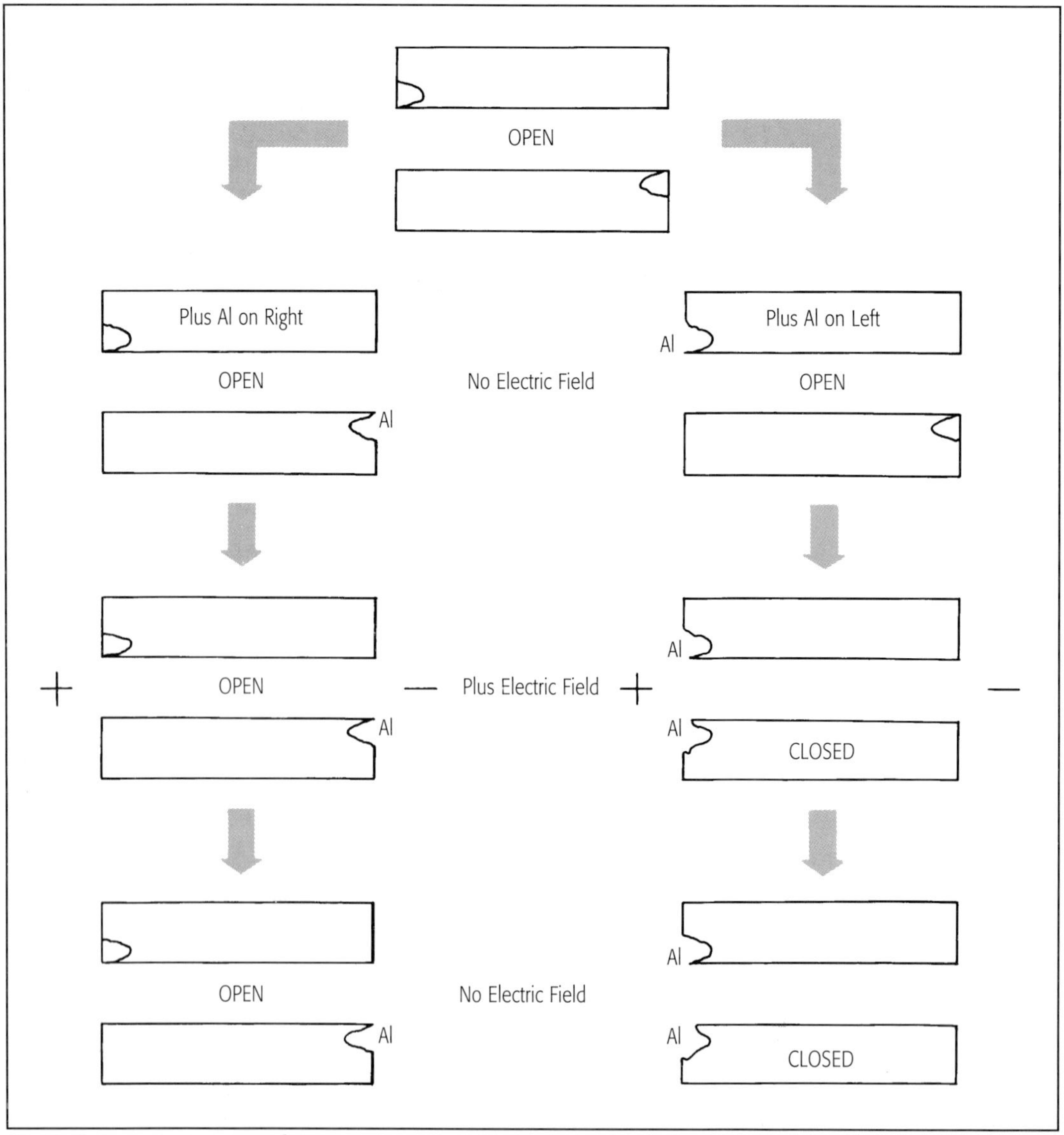

Fig. 8. Schematic drawing of a model that accounts for the asymmetric effect of aluminum. The channel is depicted in longitudinal cross-section with the shaded rectangles being the walls of the channel. The small notches on the channel represent the binding sites for aluminum. Starting at the top with an open channel, aluminum salt is added either to the right or left of the channel. This results in no effect until a membrane potential is applied. If the aluminum-containing side is made negative, the channel fails to close but closure occurs if it is made positive. When the potential is removed, the closed channel does not reopen because aluminum is bound to the site that translocated through the membrane as a result of channel closure. Reproduced with permission from Zhang & Colombini.[8]

2. Thomas L., Kocsis E., Colombini M., Erbe E., Trus B.L., Steven A.C.: *J. Structural Biol.,* **106**, (in press), 1991.
3. Mannella C.A.: *J. Cell Biol.*, **94**, 680-687, 1982.
4. Dill E.T., Holden M.J., Colombini M.: *J. Membrane Biol.*, **99**, 187-196, 1987.
5. Oortgiesen M., van Kleef R.G.D.M., Vijverberg H.P.M.: *J. Membrane Biol.*, **113**, 261-268, 1990.
6. Baes C.F. Jr., Mesmer R.E.: *The Hydrolysis of Cations*, John Wiley, New York, 1976.
7. Zhang D., Colombini M.: *Biochim. Biophys. Acta*, **991**, 68-78, 1989.
8. Zhang D., Colombini M.: *Biochim. Biophys. Acta*, **1025**, 127-134, 1990.

Aluminum and Neurodegenerative Disorders

J. Savory, M.M. Herman, C.D. Katsetos, M.R. Wills

1. Human Neurodegenerative Diseases

a) Alzheimer's Disease: Senile Plaques and Neurofibrillary Tangles

The neuropathologic hallmarks in Alzheimer's disease are the widespread accumulations of senile plaques and neurofibrillary tangles throughout the limbic system, and in parts of the cerebral isocortex and brainstem. Both neurofibrillary tangles and senile plaques are composed of fibrillar protein material which is to a large extent resistant to solubilization and thus is difficult to characterize. The elucidation of the underlying pathobiologic mechanism(s) implicated in the production of neurofibrillary lesions seen in Alzheimer's disease, will be aided by a definitive identification of the constituents of neurofibrillary tangles and senile plaques, along with an understanding of the mechanisms of their production, processing and incorporation into pathologic lesions.

Speculations as to the etiology of Alzheimer's disease include genetic factors, infectious agents and environmental toxins. Despite a large body of scientific contributions in the field, neither the etiology nor the principal pathogenesis of Alzheimer's disease have been forthcoming. One main direction of Alzheimer's disease research which would seem important, in our view, is in the experimental production, characterization and modulation of neurofibrillary lesions in animals, with a critical comparison of the latter to those found in certain human neurodegenerative disorders.

Investigations in the author's laboratory have focused on aluminum-induced neurofibrillary tangle production in the rabbit. Notwithstanding recognized differences in the antigenic characteristics between the latter and Alzheimer's disease, aluminum-induced encephalomyelopathy remains the best available experimental system for the study of neuronal cytoskeletal pathology *in vivo*. The neurofibrillary tangles in Alzheimer's disease contain by electron microscopy (after conventional fixation procedures) two 10 nm-wide filaments twisted about each other in pairs (paired helical filaments).[1] Much of the work aimed at identifying the constituent proteins of paired helical filaments has been based on epitope analysis from immunohistochemical studies. One of the problems inherent to immunohistochemistry is the inability to distinguish components of the tangle and other cross-reactive antigenic determinants which are either associated, sequestered or entrapped within this aberrant cytoskeletal structure. More definitive data in this regard may be obtained with immunoelectron microscopy[2] and molecular biological techniques.[3] Various candidate proteins of the paired helical filament are neurofilament proteins,[4-6] microtubule associated protein-2,[7] tau,[3,8] ubiquitin,[9,10] and β-amyloid peptide.[11]

b) Neurofilamentous Accumulations in Human Neurodegenerations

It is possible that a primary event common to Alzheimer's disease, amyotrophic lateral sclerosis and possibly certain other neurodegenerative diseases, is the perturbation by some agent of the neurofilamentous component of the cytoskeleton. Consequently, it is of considerable significance to note that accumulations of neurofilaments in the perikarya and proximal axons are seen in amyotrophic lateral sclerosis.[12,13] Straight filaments about 15 nm wide are found in Alzheimer's disease[14-16] and Pick's disease,[17] and these share surface epitopes with paired helical filaments and undoubtedly contain neurofilament proteins. «Inappropriate» acquisition of abnormally phosphorylated neurofilament epitopes in neuronal perikarya also has been extensively described in a variety of human neurodegenerations, including Alzheimer's disease,[18-25] Pick's disease,[26-28] progressive supranuclear palsy,[29,30] and human motoneuron disease.[31-35]

c) Aluminum Neurotoxicity

Aluminum is now known to be a toxic element and presents a major health hazard in patients with chronic renal failure, especially those on intermittent hemodialysis treatment;[36] one consequence of aluminum overload is the neurological syndrome, dialysis encephalopathy. Speculation about a relationship between aluminum neurotoxicity and Alzheimer's disease has resulted from three observations:

1) A geographical relationship between Alzheimer's disease and aluminum in drinking water has been reported.[37-39] The mental status of workers exposed to aluminum has been shown to be slightly impaired as compared to appropriate controls.[40,41]

2) Aluminum concentration in brain tissue from patients with Alzheimer's disease has been demonstrated by some workers to be elevated,[42] whereas other researchers have failed to find any changes.[43] These discrepancies are almost certainly due to the use of bulk tissue for analysis. This problem has been overcome by resorting to the microanalysis techniques of electron-probe microanalysis and laser microprobe mass spectrometry,[44] which have demonstrated aluminum deposits to be located in tangle-bearing neurons.[44,45]

3) A major argument against aluminum causing neurodegeneration in humans has been the failure to find neurofibrillary tangles in the brain of patients with dialysis encephalopathy resulting from hyperaluminemia and increased brain aluminum concentrations. However, one of us (CDK) has observed neurofibrillary tangles in 4 such patients (work in progress, 1991).

d) Cytoskeletal Protein Gene Expression

In Alzheimer's disease the RNA content in the brain is reduced, which suggests increased message degradation or reduced transcription.[46-48] Two recent studies have demonstrated a dramatic decrease in expression of the NF-L and NF-M genes in Alzheimer's disease. In one study, a decrease of 95% for NF-M and 86% for NF-L was noted in brain tissues of Alzheimer's disease as compared to controls.[49] No changes were seen for actin and tubulin. A second report shows decreases in NF-L gene expression in Alzheimer's disease which are greater than 45% over control values.[50]

2. Aluminum Binding and Biochemical Effects

a) Binding of Aluminum to Proteins

Aluminum is known to bind to many proteins, and in the process, can exert profound biochemical effects. Inhibition of protein kinase C has been reported[51,52] under poorly-defined Al^{3+} concentrations. A recent study[53] using known Al^{3+} concentrations shows that aluminum can both activate and inhibit protein kinase C; however the Al^{3+} has to be maintained at relatively high concentrations. Aluminum inhibits calpain-mediated proteolysis of neurofilament proteins. The effect appears to be an interaction with the substrate rather than a direct effect on the enzyme itself.[54]

Very little quantitative data on aluminum binding is available. Transferrin seems to be the most significant serum protein that binds aluminum; the stability constant has been determined in our laboratory, using difference spectrometry.[55] Aluminum has been shown to be bound with high affinity to NF-H.[56] Aluminum-binding to some biologically active peptides also has been reported.[57,58] Of significance to the present review is the binding of aluminum to the components of neurofilament, namely the 3 subunits: NF-L (68 kDa), NF-M (150-160 kDa) and NF-H (190-210 kDa). These proteins constitute a major component of the neuronal cytoskeleton. Aluminum undoubtedly has a high affinity for phosphate groups which must surely form the basis of binding to many proteins.[59] NF-M and NF-H are highly phosphorylated and contain 9-14 and 22-100 phosphate groups respectively.[60-62] Phosphate groups situated close to each other could enhance aluminum binding.[59]

b) Aluminum-Induced Aggregation of Neurofilament-Proteins

Most work on aluminum-induced aggregates of neurofilament-proteins has been carried out *in vivo*. However, aluminum can induce aggregates of purified neurofilament proteins *in vitro*[63, 64] and there is some indication that the highly phosphorylated proteins, NF-M and NF-H, are most susceptible to aggregation.

c) Chemistry of Aluminum

With very few exceptions, studies of aluminum toxicity and aluminum-binding have been carried out without consideration of the basic chemistry of aluminum, particularly in regard to an understanding of Al^{3+} concentrations at a neutral pH. The studies of Professor Bruce Martin have contributed significantly to an understanding of aluminum bioinorganic chemistry.[65]

3. Experimental Neurodegeneration Induced by Aluminum

a) Aluminum-Induced Neurodegeneration *in vivo*

Aluminum salts administered intracerebrally or peripherally in rabbit, cat, ferret and dog induce neurofibrillary tangles,[66-69] which consist of perikaryal accumulations of 10 nm straight filaments (neurofilaments).[70, 71] Although these neurofibrillary tangles differ ultrastructurally and, to some extent in biochemical composition from neurofibrillary tangles in Alzheimer's disease, they also share phosphorylated epitopes of the high- (NF-H) and middle- (NF-M) molecular weight isoforms of neurofilament protein.[72, 73]

Studies carried out in the authors' laboratory have provided new information on aluminum toxicity in rabbits using a water soluble aluminum compound active at a neutral pH, namely aluminum maltol. We have made a comparison of the effects of the intravenous[74] and intraventricular[75] routes of administration, which we categorize as acute, subacute and chronic aluminum treatment. The chronic model involved the placement of a permanent indwelling jugular catheter, and injection of aluminum-maltol through this catheter three times per week. Neurofibrillary tangles, immunoreactive with a monoclonal antibody to NF-H (Tp-NFP1A3, gift of Dr. V.P. Collins, Stockholm, Sweden), were observed in the oculomotor nucleus in one-third of the aluminum maltol-treated rabbits in which the nucleus could be located; the rabbits were treated for 12, 20, and 29 weeks. None of the 17 animals in either the maltol-treated or untreated control groups in which the nucleus was found, demonstrated tangles. A slight increase in brain tissue aluminum concentration was confirmed by an electrothermal atomic absorption spectrophotometric method.

We also have employed the use of the acute and subacute route of administration of aluminum maltol via the cerebral lateral ventricles. In our initial study, 9 white New Zealand rabbits were used: these were 5 males and 4 females, age 4-9 months old and weighing from 3.1-5.0 kg. The rabbits were divided into 4 experimental groups, namely 2 untreated controls, 3 short-term (acute) animals treated with aluminum maltol, 2 subacute animals treated with aluminum maltol, and 2 subacute treated with maltol. The 3 acute animals received intraventricular injections of aluminum maltol and were sacrificed 5 hours after the initial injection. Two subacutely treated animals received intraventricular injections of aluminum maltol on postoperative days 1,3 and 4, while the others received equivalent volume injections of equimolar maltol solution intraventricularly on the same injection schedule. The subacute aluminum maltol-treated animals developed widespread neurofibrillary degeneration with a predilection for motoneuron involvement. Affected neurons exhibited intense filamentous immunostaining with a panel of monoclonal antibodies against phosphorylated (SMI-31), non-phosphorylated/phosphatase sensitive (SMI-32), and dephosphorylation-independent (SMI-33) epitopes of NF-H and -M. By contrast, these lesions were nonreactive with monoclonal antibodies to microtubule associated protein-2, Class III neuron-associated β-tubulin, and a generic β-tubulin. Also observed were *intranuclear* filamentous bundles (revealed by standard electron microscopy) in large neurons from both the intraventricularly and intravenously treated animals.

Our results suggested: (i) that aluminum maltol produces similar, but, more widespread degeneration of projection-type neurons, as compared to less water-soluble aluminum compounds; (ii) that there is a distinct propensity for degeneration of large projection-type neurons, including subcortical, brainstem and spinal cord neurons; and (iii) that the aberrant polymeric aggregates in the somata and proximal neurites of affected neurons demonstrate diverse reactivities toward phosphorylation-independent neurofilament epitopes, but are negative for certain antigenic determinants of the microtubule cytoskeleton.

In addition, we followed the time course of neurofibrillary tangle formation in this rabbit system following administration of a single dose of 100 μl 25 mM aluminum maltol. In the lumbar spinal cord we observed neurofibrillary tangles as early as 36 hours after treatment.

b) Aluminum-Induced Neurodegeneration *in vitro*

Studies in the authors' laboratory [76] have led to the development of a neuronal culture system which was used to evaluate the neurotoxic effects of aluminum maltol on fetal rabbit midbrain sections containing the oculomotor nucleus. Cultures were treated with 5, 7, 9, 11, 13 and 15 μmol/l aluminum maltol or 39 and 45 μmol/l maltol (molal equivalents to 13 and 15 μmol/l aluminum maltol). Control cultures were maintained in nutrient medium alone. Silver-positive neuritic swellings and occasional perikaryal neurofibrillary tangles were observed in cultures treated with 11, 13 and 15 μmol/l aluminum maltol. The number of tangles (involved neurons) produced in aluminum maltol-treated cultures were counted and compared to (untreated) controls. We noted a total of 3,7 and 7% of involved neurons following treatment with 11, 13 and 15 μmol/l aluminum maltol respectively, and none in the control group. By immunohistochemistry, neurofibrillary tangles were immunoreactive with monoclonal antibodies SMI-31, -32 and -33 against epitopes of the NF-H and NF-M. By contrast these lesions were non-reactive with monoclonal antibodies recognizing tau, microtubule associated protein-2 and different β-tubulin isotypes (Class III β-tubulin and a generic tubulin). By electron microscopy, the perikaryal tangles consisted of focal accumulations of 10 nm straight filaments.

c) Aspects of Abnormal Phosphorylation in Aluminum-Induced Neurofibrillary Degeneration

The notion of an abnormal or «atopic» neurofilament phosphorylation in aluminum-induced neurofibrillary degeneration [77-82] is based on the conventional belief that neurofilaments normally become phosphorylated as they enter the axonal compartment and that neuronal perikaryal and proximal axons are thus devoid of phosphorylated neurofilaments.[83] The significance of this seemingly altered pattern of phosphorylation of neurofilaments in the pathogenesis of neurodegenerative states in general, and in experimental aluminum-induced neurofibrillary degeneration in particular, remains unclear. It is known that the least phosphorylated form of neurofilament is present in the moving SCa wave of axonally transported protein, whereas the highly phosphorylated neurofilament form(s) remains relatively stationary.[84] Although significant neurofilament phosphorylation occurs in the perikaryon shortly after synthesis, additional phosphate is added during axonal transport.[84] The accumulation of neurofilament polymer in the somata of neurons associated with aluminum-induced neurodegeneration, may represent a stationary neurofilament assemblage resulting from abnormal phosphorylation. The identification of a conserved phosphorylation site in mammalian neurofilaments (recognized by a class of monoclonal antibodies defined as type A) [85] and evidence that neurofibrillary tangles of Alzheimer's disease, microtubule associated protein-2, and tau (and perhaps other neuronal phosphoproteins) share sequences containing phosphorylation repeats identical to those present in NF-M and NF-H,[85] has augmented our awareness about the presence of such common epitopes in otherwise dissimilar neurofibrillary lesions. Hence, the recognition of Alzheimer's disease neurofibrillary tangles and aluminum-induced neurofibrillary lesions by SMI-31, a type A monoclonal antibody[85] against a phosphorylated NF-M and phosphorylated NF-H, perhaps offers two different interpretations; i.e., in Alzheimer's disease tangles, SMI-31 recognizes a cross-reactive antigenic determinant common to a number of cytoskeletal proteins in an intrinsically abnormal cytoskeletal structure (paired helical filament), whilst in aluminum-induced neurofibrillary degeneration, it reacts with phosphorylated epitopes in retained masses of neurofilaments, otherwise exhibiting diverse anti-neurofilament immunoreactivities. Of importance also, are quantitative data in rats administered aluminum in the drinking water, showing 150-200% increases in NF-H phosphorylation in the cerebral cortex and brainstem.[86]

d) Aberrant Perikaryal Neurofilament Polymerization and Aldehyde Fixation-Related Immunohistochemical Artifacts

It has been recently demonstrated that monoclonal antibodies against phosphorylated neurofilaments (including SMI-31, which ordinarily stains axons in aldehyde-fixed specimens), are immunoreactive with neuronal perikarya in fresh frozen mammalian brain sections *in situ*.[87] These workers postulated that normally present phosphorylated forms of perikaryal neurofilament protein are susceptible to denaturation during ordinary tissue processing and chemical fixation. By contrast, axonal phosphorylated neurofilaments maintain their immunoreactivity even after relatively prolonged fixation. The explanation for this intriguing characteristic of phosphorylated neurofilament in axons may rest with their aligned conformation (perhaps on the context of neurofilament polymerization) which confers resistance to denaturation and hence preservation of antigenicity.[88] Furthermore, soluble, phosphorylated forms of NF-H have been shown in the perikaryal of cultured neuronal cells (fixed in 4% paraformaldehyde) wherein they are present as oligomers in the Triton-soluble fraction.[88] Thus it is plausible that *soluble* phosphorylated neurofilament, normally present in the perikaryon in a monomeric (and/or oligomeric) form, loses its antigenicity following aldehyde fixation and delipidation, whereas *polymerized* phosphorylated neurofilament protein, normally present in the axon, retains it by virtue of a distinct conformation. If polymerization of neurofilament is indeed a factor of antigenic preservation (i.e., under conventional conditions of aldehyde based fixation), this would partly explain the consistent observation that anti-phosphorylated neurofilament reactivity is unimpaired in the perikaryal polymeric neurofilamentous masses. Further, it should be pointed out that the somata of certain neurons, e.g., the mesencephalic nucleus of the trigeminal[89] and also pyramidal neurons of striate and peri- and para- striate isocortical ribbon in untreated and maltol-treated, formalin perfused, animals in one of our studies,[74] display *normal* phosphorylated NF-H-like reactivity in their perikaryal. The association of neurofilament polymerization and antigenic preservation of neurofilament proteins after aldehyde fixation is further inferred by the following observation: whereas the antigenicity of certain non-phosphorylated neurofilament epitopes was severely impaired after formalin fixation (evidenced as abolition of the normally expected perikaryal staining), the staining in the aluminum-induced lesions with anti-nonphosphorylated phosphorylation-dependent and phosphorylation independent NF-H/M monoclonal antibodies remained remarkably intense, suggesting that a conformational change possibly connected with neurofilament polymerization might have accounted for this phenomenon.

e) Expression of Non-phosphorylated, Phosphorylation Dependent and Independent Neurofilament Epitopes in Aluminum-Induced Neurofibrillary Degeneration

In addition to the presence of phosphorylated neurofibrillary degeneration epitopes in lesions of aluminum-induced neurofibrillary degeneration, noteworthy is the presence of non-phosphorylated and phosphate independent neurofilament epitopes as revealed by immunoreactivity of these lesions with monoclonal antibodies SMI-32 and SMI-33. It has been reported by other investigators that neurofibrillary lesions induced by aluminum are recognized by monoclonal antibodies to non-phosphorylated neurofilaments [78] and by SMI-33.[85]

f) Aluminum Effects on Gene Expression

Recently Muma et al.[90] using intraventricular administration of aluminum lactate to rabbits have shown in spinal cord motor neurons an approximately 3-fold reduction in the expression of genes coding for NF-L and tubulin; cDNA probes were used for both *in situ* hybridization and Northern blot analysis. Different possible explanations for these observations can be proposed, such as aluminum cross-linking histone Hl to DNA,[91] while another is that reduced expression of NF-L could occur secondary to aluminum-induced alterations. Muma et al.[90] allowed the rabbits to demonstrate neurotoxic effects before sacrificing them and analyzing the tissue. Recently,[92] a more extensive study was carried out on weanling rabbits injected intracisternally with aluminum chloride and sacrificed at 1, 2, 7 and 18 days. Northern analysis and *in situ* hybridization were used to evaluate gene expression for NF-L, NF-M, NF-H, neuronal specific enolase, glial fibrillary acidic protein and amyloid precursor protein. In all cases except for glial fibrillary acidic protein, there was a decrease in gene expression which occurred early and concurrently with the development of neurofibrillary lesions. These investigators suggested that the accumulation of neurofibrillary tangles resulting from direct injection of an aluminum compound into the central nervous system is not due to a selective effect on neurofilament gene expression. The fact that glial fibrillary acidic protein message levels are not reduced indicates strongly that aluminum exerts its effect on neurons but not on glia.

Conclusions

Aluminum-induced neurodegeneration, both in experimental animals and *in vitro*, should continue to be a useful system to study the aggregation of cytoskeletal proteins in neuronal perikaryal. Whether the mechanism will be similar to or totally different from that seen in certain human neurodegenerative diseases remains to be elucidated. It does appear that some interaction with, or effect on phosphorylated cytoskeletal proteins constitutes a plausible primary event in the aluminum-induced (and possibly also in some of the human) neurodegenerations. Studies on aluminum-induced neurotoxicity may, therefore, aid in elucidating the underlying pathobiologic mechanisms in Alzheimer's disease, amyotrophic lateral sclerosis and other related human neurodegenerative disorders.

References

1. Wischik C.M., Crowther R.A., Stewart M., Roth M.: *J. Cell. Biol.*, **100**, 1905-1912, 1985.
2. Wischik C.M., Novak M., Edwards P.C., Klug A., Tichelaar W., Crowther R.A.: *PNAS*, **85**, 4884-4888, 1988.
3. Goedert M., Wischik C.M., Crowther R.A., Walter J.E., Klug A.: *PNAS*, **85**, 4051-4055, 1988.
4. Dahl D., Selkoe D.J., Pero R.T., Bignami A.: *J. Neurosci.*, **2**, 113-119, 1982.
5. Gambetti P., Velasco M.E., Dahl D., Bignami A., Roessmann U., Sindely S.D.: *Aging of the Brain and Dementia*, **13**, 55-63, 1980.
6. Miller C.C.J., Brion J.P., Calvert R., Chin R.K., Eagles P.A.M., Downes M.J., Flament-Duran J., Haugh M., Kahn J., Probst A., Ulrich J., Anderton B.H.: *EMBO J.*, **5**, 269-276, 1986.
7. Kosik K.S., Duffy L.K., Dowling M.M., Abraham C., McCluskey A., Selkoe D.J.: *PNAS*, **81**, 7941-7945, 1984.
8. Wood J.G., Mirra S.S., Pollock N.J., Binder L.I.: *PNAS*, **83**, 4040-4043, 1986.
9. Mori H., Kondo J., Ihara Y.: *Science.*, **235**, 1641-1644, 1987.
10. Manetto V., Perry G., Tabaton M., Mulvihill P., Fried V.A., Smith H.T., Gambetti P., Autilio-Gambetti L.: *PNAS*, **85**, 4505-4105, 1987.
11. Guiroy D.C., Miyazaki M., Multhaup G., Fischer P., Garruto R.M., Beyreuther K., Masters C.L., Simms G., Gibbs C.J., Gajdusek D.C.: *PNAS*, **84**, 2073-2077, 1987.

12. Hirano A., Donnenfeld H., Sasaki S., Nakano I.: *J. Neuropathol. Exp. Neurol.*, **43**, 461-470, 1984.
13. Sobue G., Hashizume Y., Yasuda T., Mukai E., Kumagai T., Mitsuma T., Trojanowski J.Q.: *Acta Neuropathol.*, **79**, 402-408, 1990.
14. Yagishita S., Itoh Y., Nan W., Amano N.: *Acta Neuropathol.*, **54**, 239-246, 1981.
15. Metuzals J., Montpetit V., Clapin D.F.: *Cell Tissue Res.*, **214**, 455-482, 1981.
16. Crowther R.A.: *PNAS*, **88**, 2288-2292, 1991.
17. Perry G., Stewart D., Friedman R., Manetto V., Autilio-Gambetti L., Gambetti P.: *Am. J. Pathol.*, **127**, 559-568, 1987.
18. Anderton B.H., Breinberg D., Downes M.J., Green P.J., Tomlinson B.E., Ulrich J., Wood J.N., Khan J.: *Nature*, **298**, 84-86, 1982.
19. Rasool C.G., Abraham C., Anderton B.H., Haugh M., Kahn J., Selkoe D.J.: *Brain Res.*, **310**, 249-260, 1984.
20. Gambetti P., Autilio-Gambetti L., Perry G., Shecket G., Crane R.C.: *Lab. Invest.*, **49**, 430-435, 1983.
21. Sternberger N.H., Sternberger L.A., Ulrich J.: *PNAS*, **82**, 4274-4276, 1985.
22. Cork L.C., Sternberger N.H., Sternberger L.A., Casanova M.F., Stuble R.G., Price D.L.: *J. Neuropathol. Exp. Neurol.*, **45**, 56-64, 1986.
23. Lee V.M.-Y., Otuos L., Schmidt M.L., Trojanowski J.Q.: *PNAS*, **85**, 7384-7388, 1988.
24. Schmidt M.L., Gur R.E., Gur R.C., Trojanowski J.Q.: *Ann. Neurol.*, **23**, 184-189, 1988.
25. Schmidt M.L., Lee V.M.-Y., Trojanowski J.Q.: *Lab. Invest.*, **60**, 513-522, 1989.
26. Rasool C.G., Selkoe D.J.: *N. Eng. J. Med.*, **812**, 700-705, 1985.
27. Perry G., Stewart D., Friedman R., Manetto V., Autilio-Gambetti L., Gambetti P.: *Am. J. Pathol.*, **127**, 559-568, 1987.
28. Dickson D.W., Suzuki K., Davies P.: *J. Neuropathol. Exp. Neurol.*, **45**, 380, 1987 (Abstr.).
29. Galloway P.G.: *Neurosci. Lettters*, **91**, 148-153, 1988.
30. Schmidt M.L., Lee V.M.-Y., Hurtig H., Trojanowski J.Q.: *Lab. Invest.*, **59**, 460-466, 1988.
31. Schmidt M.L., Carden M.J., Lee V.M.-Y., Trojanowski J.Q.: *Lab. Invest.*, **56**, 282-294, 1987.
32. Manetto L., Sternberger N.H., Perry G., Sternberger L.A., Gambetti P.: *J. Neuropathol. Exp. Neurol.*, **47**, 642-653, 1988.
33. Munoz D.G., Greene C., Perl D.P., Selkoe D.J.: *J. Neuropathol. Exp. Neurol.*, **47**, 9-18, 1988.
34. Leigh P.N., Anderton B.H., Dodson A., Gallo J.M., Swash M., Power D.M.: *Neurosci. Letters*, **93**, 197-203, 1988.
35. Shankar S.K., Yanagihara R., Garruto R.M., Grundke-Iqbal L., Kosik K.S., Gajdusek D.C.: *Ann. Neurol.*, **25**, 146-151, 1989.
36. Wills M.R., Savory J.: *Lancet*, **ii**, 29-34, 1983.
37. Martyn C.N., Osmond C., Edwardson J.A., Barker D.J.P., Harris E.C., Lacey R.F.: *Lancet* (Jan. 14), 59-62, 1989.
38. Flaten T.P.: Thesis no. 51. Institutt Fort Norganisk Kjemu, Norges Tekniske Hogskole, Trondheim, 1986.
39. Vogt T.: Sostale og Okonomiske Studier 61: 1. Oslo: Central Bureau of Statistics of Norway, 1986.
40. Hosovski E., Mastelica Z., Sunderic D., Radulovic D.: *Med. Lav.*, **2**, 119-123, 1990.
41. Rifat S.L., Eastwood M.R., Crapper-McLachlan D.R., Corey P.N.: *Lancet*, **336**, 1162-1165, 1990.
42. Crapper D.R., Krishnan S.S., Dalton A.J.: *Science*, **180**, 511-513, 1973.
43. McDermott J.R., Smith A.I., Iqbal K., Wisniewski H.M: *Lancet*, 710-711, 1977.
44. Perl D.P., Brody A.R.: *Science*, **208**, 297-299, 1980.
45. Candy J.M., Klinowski J., Perry R.H., Perry E.K., Fairbairn A., Oakley A.E., Carpenter R.A., Atack J.R., Blessed G., Edwardson J.A.: *Lancet*, **i**, 354-356, 1986.
46. Uemura E., Hartmann H.A.: *Brain. Res. Bull.*, **4**, 301-305, 1979.
47. Mann D.M.A., Neary D., Yates P.O., Lincoln J., Snowden S.S., Starworth P.: *Neurol. Neurosurg. Psych.*, **44**, 97-102, 1981.
48. Sajdel-Sulkowska E., Marotta C.A.: *Science*, **225**, 947-949, 1984.
49. Kittur S., Hoh J., Kawas C., Tourtellotte W., Marksberry W., Adler W.: *Neurobiol. Aging.*, **11**, 285-286, 1990.
50. Somerville M.J., McLachlan D.R.C., Bergeron C., Yoong L.K.K., Percy M.E.: *Neurobiol. Aging.*, **11**, 286, 1990.
51. Katsuyama H., Kiyofumi S., Inoue Y., Sumino K.: *Arch. Toxicol.*, **63**, 474-478, 1989.
52. Cochran M., Elliott D.C., Brennan P., Chawtur V.: *Clinica Chimica Acta*, **194**, 167-172, 1990.
53. MacDonald T.: Personal communication.
54. Nixon R.A., Clarke J.F., Logvinenko K.B., Tan M.K.H., Hoult M., Grynspan F.: *J. Neurochem.*, **55**, 1950-1959, 1990.
55. Martin R.B., Savory J., Brown S., Bertholf R.L., Wills M.R.: *Clin. Chem.*, **33**, 405-407, 1987.
56. Pierson K.B., Evenson M.A.: *Biochem. Biophys. Res. Comm.*, **152**, 598-604, 1988.
57. Mazarguil H., Haran R., Laussac J.-P.: *Biochim. Biophys. Acta*, **717**, 465-472, 1982.
58. Siegel N., Haug A.: *Biochim. Biophys. Acta*, **744**, 36-45, 1983.
59. Panchalingam K., Post J.F.M., Jiminez R., Pettegrew J.W.: *Society of Biological Psychiatry*, Forty-Second Annual Scientific Program, p. 228, 1987.
60. Jones S.M., Williams R.C.: *J. Biol. Chem.*, **257**, 9902-9905, 1982.
61. Julien J.-P., Mushynski W.E.: *J. Biol. Chem.*, **257**, 10467-10470, 1982.
62. Geisler N., Kaufman E., Fischer S., Plessman U., Weber K.: *EMBO J.*, **2**, 1295-1302, 1983.

63. Diaz-Nido J., Avila J.: *Neuroscience Letters*, **10**, 221-226, 1990.
64. Troncoso J.C., March J.L., Haner M., Aebi U.: *J. Struct. Biol.*, **103**, 2-12, 1990.
65. Martin B.B.: *Clin. Chem.*, **32**, 1797-1802, 1986.
66. Klatzo I., Wisniewski H., Streicher E.: *J. Neuropathol. Exp. Neurol.*, **24**, 187-199, 1965.
67. Wisniewski H.M., Narkiewicz O., Wisniewski K.: *Acta Neuropathol.*, **9**, 127-133, 1967.
68. Yates C.M., Gordon A., Wilson H.: *Neuropathol. Appl. Neurobiol.*, **2**, 131-144, 1976.
69. Wisniewski H.M., Sturman J.A., Shek J.W.: *Acta Neurol.*, **8**, 479-490, 1980.
70. Terry R.D., Pena C., *J. Neuropathol. Exp. Neurol.*, **24**, 200-210, 1965.
71. Dahl D., Bignami A.: *Exp. Neurol.*, **58**, 74-80, 1978.
72. Sternberger N.H., Sternberger L.A., Ulrich J.: *PNAS*, **82**, 4274-4276, 1985.
73. Cork L.C., Sternberger N.H., Sternberger L.A., Casanova M.F., Struble R.G., Price D.L.: *J. Neuropathol. Exp. Neurol.*, **45**, 56-64, 1986.
74. Bertholf R.L., Herman M.M., Savory J., Carpenter R.M., Sturgill B.C., Katsetos C.D., Vandenberg S.R., Wills M.R.: *Toxicol. Appl. Pharmacol.*, **98**, 58-74, 1989.
75. Katsetos C.D., Savory J., Herman M.M., Carpenter R.M., Frankfurter A., Hewitt C.D., Wills M.R.: *Neuropathol. Appl. Neurobiol.*, **16**, 511-528, 1990.
76. Hewitt C.D., Herman M.M., Lopes M.B.S., Savory J., Wills M.R.: *Neuropathol. Appl. Neurobiol.*, **17**, 47-60, 1991.
77. Bizzi A., Gambetti P.: *Acta Neuropathol.*, **71**, 154-158, 1980.
78. Munoz-Garcia D., Pendlebury W.W., Kessler J.B., Perl D.: *Acta Neuropathol.*, **70**, 243-248, 1986.
79. Troncoso J.C., Sternberger N.H., Sternberger L.A., Hoffman P.N., Price D.L.: *Brain Res.*, **364**, 295-300, 1986.
80. Pendlebury W.W., Beal M.F., Kowall N.W., Solomon P.R.: *J. Neurol. Transm.* (Suppl.), **24**, 213, 1987.
81. Pendlebury W.W., Beal M.F., Kowall N.W., Solomon P.R.: *Neurotoxicol.*, **9**, 503-510, 1988.
82. Kowall N.W., Pendlebury W.W., Kessler J.B., Perl D.P., Beal M.F.: *Neuroscience*, **29**, 329-337, 1989.
83. Sternberger L.A., Sternberger N.H.: *PNAS*, **80**, 6126-6130, 1983.
84. Hollenbeck P.J.: *J. Cell. Biol.*, **108**, 223-227, 1989.
85. Lee V.N.-Y., Otvos L. Jr., Carden M.J., Hollosi M., Dietzschold B., Lazzarini R.A.: *PNAS*, **85**, 1998-2002, 1988.
86. Johnson G.V.W., Jope R.S.: *Brain Res.*, **456**, 95-103, 1988.
87. Poltorak M., Freed W.J.: *Brain Res.*, **480**, 349-354, 1989.
88. Shea T.B., Majocha R.E., Marotta C.A., Nicon R.A.: *Neurosci. Letters*, **92**, 291-297, 1988.
89. Poltorak M., Freed W.J.: *Exp. Neurol.*, **97**, 735-738, 1987.
90. Muma N.A., Troncoso J.C., Hoffman P.N., Koo E.H., Price D.L.: *Mol. Brain Res.*, **3**, 115-122, 1988.
91. McLachlan D.R.C., Kruck T.P.A., Lukin W.J.: *Neurobiol. Aging*, **11**, 315, 1990 (Abstr).
92. Parhad I.M., Krekowski C.A., Mathew A., Tran P.M.: *Cell. Mol. Neurobiol.*, **9**, 123-138, 1989.

The Toxicity of Aluminium and the Effect of Silicon on its Bioavailability

J.D. Birchall

Abstract

When aluminium gains ingress into living organisms, toxic responses follow at relatively small concentrations of the element. Aluminium toxicity is well documented in plants, fish, higher animals and in man. A major mechanism of cellular toxicity is the ability of aluminium to displace magnesium and calcium at key sites at which these metal ions are catalytic. Plants in acid soils with available aluminium present in the interstitial water show stunted root growth; fish in acidic water containing < 10 micromoles/l aluminium show damage to gill structure and function leading to death; renal dialysis patients with plasma aluminium levels < 10 micromoles/l display anaemia, bone disorders and progressive cognitive degeneration leading to dementia and death. The concept of aluminium as a neutral element – neither essential nor damaging – is no longer tenable. Two important questions are raised. The first relates to the mechanism of toxicity at the cellular level. This paper addresses a perhaps more fundamental question. With aluminium being the third most abundant element in the earth's crust after oxygen and environmentally ubiquitous, how did biology evolve at all? What determines the availability of aluminium to living organisms?

In this paper it is argued that the bioavailability of aluminium is controlled by the concentration of silicon and, indeed, that this is the major role for silicon rather than any intrinsic essentially as hitherto proposed. The balance of these two elements in the environment is thus important and, in particular, cannot be neglected in any consideration of the epidemiology of aluminium-related disease.

Silicon – An Essential Element?

Silicon in the Geosphere

Silicon, the second most abundant element in the earth's crust after oxygen, is the sixth most abundant element in the biosphere, being present in trace amounts in all organisms and averaging 0.033 atomic percent in the biosphere as a whole. In rocks and soil minerals it is present associated with oxygen as the oxide, SiO_2, and the silicates are based on $[SiO_4]^{4-}$ tetrahedra either isolated (as in the ortho-silicates) or as polymers in which tetrahedra, joined together at corners via oxygen, are arranged in chains, sheet or three-dimensional networks. The varied structures of the silicate minerals result from such arrangements of tetrahedra, from the ability of the $[AlO_4]^{5-}$ tetrahedra to replace $[SiO_4]^{4-}$ and from the need for additional cations in the structures so formed to maintain charge neutrality. When silica or silicates are in contact with water there is invariably some hydrolysis of Si-O-Si bonds and silicic acid is released into the aqueous phase. Silicic acid, $Si(OH)_4$, is a weak acid with $Si(OH)_3O^-$ being formed at pH 9.4 and $Si(OH)_2O_2^{2-}$ at pH 12.4.

Thus in the biologically relevant environment the element silicon exists as the neutral $Si(OH)_4$. This has a solubility of 100-140 ppm in pure water and above this level polycondensation occurs with the formation of hydrated silica (opaline silica, $SiO_2 \cdot nH_2O$). This is found in the extremities of grasses as opaline phytoliths. The silicon absorbed by organisms is invariably in the form of silicic acid. (For a review see reference 1).

Silicon in Biology

The fact that silicon is invariably present in small concentrations in all living organisms has prompted questions as to a possible functional role. It is regarded as being essential for growth and development in the diatom, not only for the construction of the siliceous frustule (Fig. 1) but also for major metabolic processes. Cells of the diatom Navicula Pelliculosa starved of silicon show reduced division and inhibition of net protein production, DNA synthesis and chorophyll synthesis – conditions reversed when sufficient silicon is added.[2,3] Some plants have been claimed to require silicon, in particular rice.[4]

The experiments which caused silicon to be listed as an element essential for the normal growth and development of higher animals were carried out in 1972 and reported independently by Schwarz[5] and Carlisle.[6] These workers reared rats and chicks respectively on a silicon-deficient diet within a controlled environment. In both species daily weight gains were reduced and could be recovered by silicon (as silicate) supplementation to the diet. In both species there were marked changes to the development and structure of collagenous connective tissue and to bone.

In chicks, for example, the leg bones were shortened in silicon deficiency; relatively flexible metatarsal bones and fragile femur and tibia were notable. In the deficient birds, the beak was flexible, the comb poorly developed and the wattles absent. In the rat the structure of the skull was altered by silicon deficiency. These findings suggested a key role for silicon in osteogenesis and this was supported by the observation (electron probe microanalysis) of silicon located at the osteoid layer cement line in normal tibia. About 0.5% Si was present within this narrow region whereas low levels were detected in adjacent periosteal regions and in fully mineralised bone.[7] Another finding was that of altered cartilage composition with reduced collagen content. Initially, it was considered that silicon was bound in the collagen and mucopolysaccharidic components of connective tissue[8] in a form not reactive to molybdate (which reacts with ortho-silicic acid), stable to 8 molar urea (and thus not hydrogen bonded) and not liberated by autoclaving or treatment with weak acid or alkali. It was considered that this strongly bound silicon (200-400 micrograms/gram) could act as a labile cross-link via Si-O-Si, promoting association between protein and polysaccharide components and affecting the tertiary structure and stability of connective tissue. This view was later discarded as improved analytical methods progressively reduced the levels of silicon found.[9]

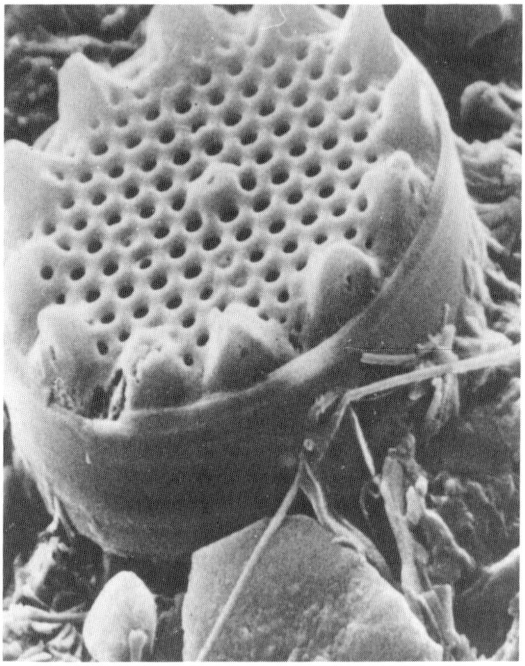

Fig. 1. The siliceous frustule of the diatom.

Significant levels of silicon were reported to be present in various tissues, notably in skin and in the aorta. In the human aorta, the Si level was reported to fall with age (from 205 microgram Si/100mg N in infants to 86 micrograms/100mg N at age 30-40) and also to fall with increasing degrees of sclerotic damage and calcification.[10]

Since 1972 there has been a search for the biochemical mechanisms that may be involved in the presumed essentially of silicon. A group of scientists meeting in 1986 in a Ciba

Foundation Conference on « Silicon Biochemistry »[11] concluded that there was no evidence for Si-C bonds in biology, little or no evidence for the formation of stable Si-O-C bonds and no convincing evidence of the stable binding of silicon (as silicic acid) to any biological molecules or macromolecules. How then does silicon deficiency produce evident pathological changes? It was suggested at that meeting that the only chemistry possible under physiological conditions with silicic acid would be with metal ions, trivalent iron being a candidate (Birchall and Espey reference 11). Being a very weak acid, silicic acid interacts only with basic cations and at a pH just below that at which the hydroxide is precipitated. Thus at physiological pH, interaction would be possible with ferric iron and with aluminium but not with calcium or magnesium which become basic above pH 9. There are indeed interactions with ferric iron. At a 3:1 molar ratio of Si (as silicic acid) to Fe, the precipitation of ferric hydroxide is inhibited. This is not due to complex formation but to the formation of particles of hydrated ferric oxide, 10-15nm diameter, stabilised against aggregation and precipitation by a coating of silicic acid. The formation of Fe-O-Si bonds is too slow to prevent the olation and polymerisation of Fe-O-Fe species. This is in contrast with aluminium: there are specific interactions between hydroxyaluminium species and silicic acid discussed in detail later.

The Involvement of Aluminium – The First Clue

Carlisle,[12] noting the effect of silicon deficiency on collagen formation, investigated the activity of prolyl hydroxylase in cartilage tissue from Si-deficient chicks. It was found that the activity of this enzyme was low as measured by hydroxyproline production. Activity was increased on the addition of silicic acid to the culture and it was concluded that Si was in some way a « co-factor » in the operation of this enzyme, the known co-factors being Fe, ascorbate, oxygen and 2-oxyglutarate. It is difficult to envisage any chemistry between silicic acid and any of these co-factors or with the enzyme protein.

In unrelated work, it had been noted that the activity of hexokinase was low and could be greatly increased by the addition of citrate.[13] This was due to contamination of the system by aluminium which binds strongly to ATP in place of the essential magnesium but which is removed by complexation with citrate. Could aluminium, in silicon-deficient conditions, be « poisoning » the prolyl hydroxylase system investigated by Carlisle? This has been investigated. (Birchall and Espey in reference 11, p. 140-159).

The effect of aluminium and silicic acid on the activity of prolyl hydroxylase was determined by measuring the release of ^3H as ^3H_2O from ^3H-prolyl protocollagen which accompanies the hydroxylation of proline at position 4. The results are given below.

The effect of aluminium and silicic acid on the activity of prolyl hydroxylase.

Addition 100 micromole (all other co-factors present)	% Inhibition of control
$Si(OH)_4$	0
Zn	100
Fe then Al	20
Al then Fe	55
Al plus 6 micromole $Si(OH)_4$ then Fe	0

The effect of the order of addition of iron and aluminium on activity indicates that aluminium competes weakly with iron for binding in the enzyme. The binding of aluminium is completely inhibited when silicic acid is present, suggesting a selective interaction between aluminium and silicic acid. This has been investigated.

The Formation of Hydroxyaluminosilicates

The uncharged species $Si(OH)_4$ interacts with metal ions only when the ion is basic, i.e. when M-OH has replaced M-H_2O, which oc-

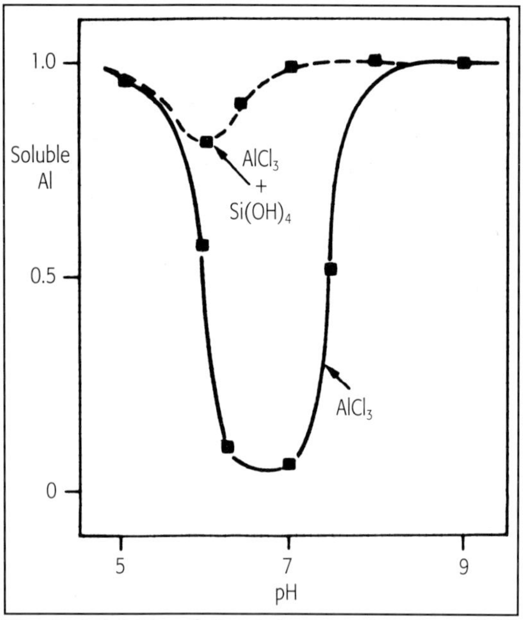

Fig. 2. Solubility of aluminium as a function of pH with and without silicic acid (20 °C). Solutions contained 0.1 mMl^{-1} $AlCl_3$ with and without 0.5 mMl^{-1} $Si(OH)_4$.

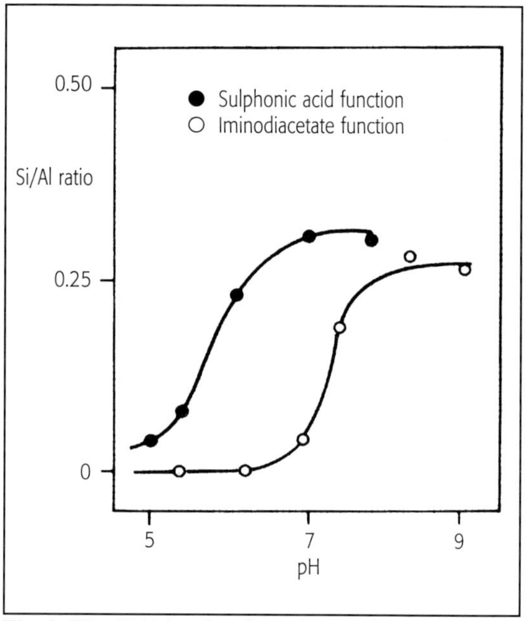

Fig. 3. The Si/Al ratio of species retained on cation-exchange resins — sulphonate and iminodiacetate — as a function of pH. Solutions contained 0.1 mMl^{-1} $AlCl_3$ with and without 0.5 mMl^{-1} $Si(OH)_4$.

curs at lower than pH 5 in the case of aluminium. The mutual affinity between aluminium ions and silicic acid is a unique one in inorganic chemistry as pointed out by Iler.[14] The hexaaquoaluminium cation, $[Al(H_2O)_6]^{3+}$, is stable only below pH 3 and above this hydrolysis occurs with the progressive formation of $[Al(H_2O)_5(OH)]^{2+}$, $[Al(H_2O)_4(OH)_2]^{+}$, the precipitation of $Al(OH)_3$ at pH 6.2 and increasing solubility at more alkaline pH due to the formation of the aluminate anion, $Al(OH)_4^-$. Figure 2 shows the solubility of aluminium as a function of pH with and without silicic acid present. It appears that solubility is increased by silicic acid. If such solutions are heated a solid phase is eventually formed in which the Si/Al ratio is 0.5; this solid is imogolite, a tubiform structure of ideal composition $(OH)_3Al_2O_3Si(OH)$ which can be considered as a single $Al(OH)_3$ Gibbsite sheet with the inner surface hydroxyls replaced by silicic acid.

Solutions held at low temperature have been shown, by the infra-red examination of solids recovered from solutions by freeze-drying[15] and by ion exchange experiments,[16] to contain precursors to or fragments of this structure.

Solutions containing 0.1 mM aluminium (as chloride) with and without 0.5 mM silicic acid per litre of 0.01 M NaCl solution were prepared. No filterable solids formed for some weeks at room temperature. When solids did form these had an Si/Al ratio in the range 0.3 to 0.6, consistent with protoimogolite species.

Solutions 20 hours old and at various pH values were passed through a cation exchange resin having sulphonate functionality (Amberlite 252). The Si/Al ratio in species retained on the resin was determined by acid washing and analysis of the extract. The results are shown in figure 3. Clearly, hydroxy-aluminosilicate species are formed, from pH values above 5, long before the formation of solids that can be recovered on membrane filters. Such species form even in the presence of citrate, an aluminium chelator. Solutions containing 0.1 mM/l aluminium and equimolar (0.5 mM/l) concentrations of silicic acid and citrate at pH

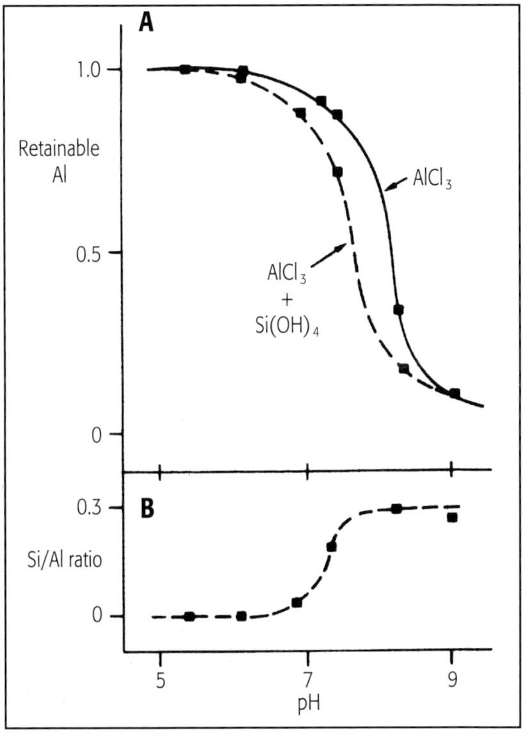

Fig. 4 A-B. A) Fraction of aluminium retained on an iminodiacetate resin as a function of pH with and without silicic acid. Solutions contained 0.1 mMl^{-1} $AlCl_3$ with and without 0.5 mMl^{-1} $Si(OH)_4$. B) Si/Al ratio of the retained species.

7.4 yielded no filterable solids after 12 weeks. However, ion-exchange experiments as above on 20 hour old solutions revealed the presence of hydroxy-aluminosilicates having an Si/Al ratio of < 0.5. Hydroxy- alumino silicate species are formed in the presence of citrate although their aggregation to filterable size is greatly retarded by citrate. This was studied further by passing solutions containing aluminium and silicic acid over a column containing an iminodiacetate-functional resin. The solutions contained 0.1 mM/l aluminium (as chloride) with and without silicic acid at 0.5 mM/l in a background of 10 mM/l sodium bicarbonate. The results are shown in figure 4. The drop in aluminium retention is shifted to a lower pH when silicic acid is present. This is the result of a change in aluminium coordination from carboxylate to silicate at about pH 7. Figure 4 also shows the Si/Al ratio of the retained species.

The conclusion is that hydroxy-aluminosilicates species are formed in pure solutions from about pH < 5 upwards and from about pH 7 in solutions containing an aluminium chelator such as citrate. These species correspond to imogolite precursors (protoimogolite), having an Si/Al ratio of 0.25 to 0.5. They are slow to grow to filterable size and aggregation is further retarded in the presence of citrate. There are no similar reactions with ferric iron and it is selective hydroxy-aluminosilicate formation that prevented the inhibition of proly hydroxylase in the experiment reported above.

It is clearly important to know the effect of silicic acid concentration on the formation of hydroxy-aluminosilicates. Is the concentration within the range found in the environment or in physiology? Figure 5 shows the Si/ Al ratio in hydroxy-aluminosilicates phases as a function of silicic acid concentration at physiological pH. Clearly, species with low Si/Al ratio form at very low silicic acid concentrations but the imogolite-precursor species (which will be the most stable) require at least 100 micromole/l to form.

Although the growth rate of sub-colloidal hydroxy-aluminosilicates species is slow (slower than aluminium hydroxide), their ultimate solubility is significantly lower than that of Gibbsite. The precipitation of Gibbsite lowers the level of aluminium in solution to 10^{-7} M/l whilst the formation of imogolite reduces it to 10^{-11} M/l. The formation of imogolite-precursor species with Si/Al ratio 0.5 may be regarded as the adsorption of silicic acid on a preformed, primitive Gibbsite nucleus, retarding the continued growth of Gibbsite and producing instead a poorly crytalline solid. A mechanism similar to this appears to be the case in the formation of calcium silicate hydrate, a poorly crystalline solid of variable stoichiometry formed when silicic acid (disilicate anions since the pH in this case is > 12) adsorb onto a preformed calcium hydroxide nucleus.[17]

What are the consequences as regards the

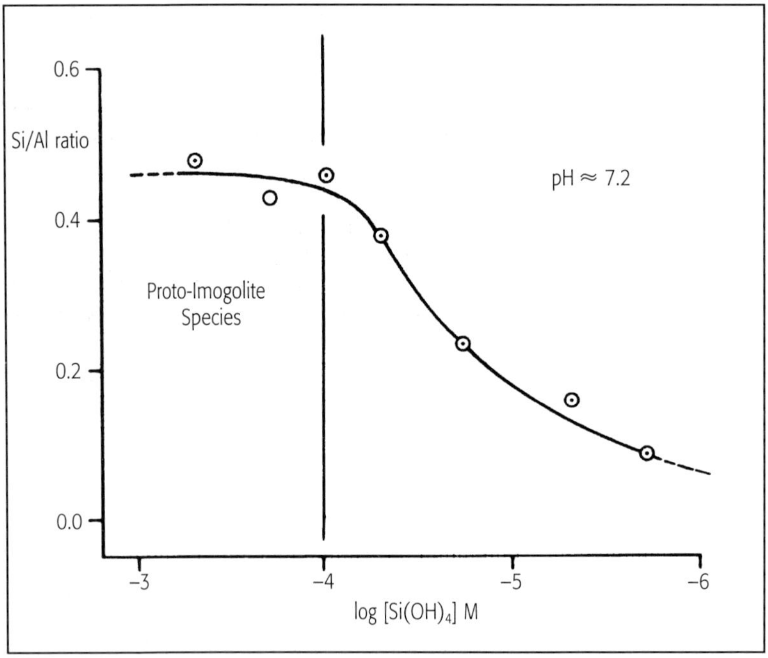

Fig. 5. Effect of silicic acid concentration on the Si/Al ratio of hydroxy-aluminosilicate species formed at pH 7.2, 20 °C.

bioavailability of aluminium of the formation of hydroxy-aluminosilicates?

The Effect of Silicic Acid on the Toxicity of Aluminium to Fish

The toxic effect of aluminium to fish in acid water is well documented.[18, 19, 20] Death is due to an effect of aluminium on gill structure and function with loss of iono-regulatory and osmo-regulatory function. Organic constituents of water such as humic acid ameliorate toxicity by binding aluminium more strongly than binding sites on the gill surface.[21] Silicic acid is a component of river and lake water but the question as to its effect on the toxicity of aluminium had not been investigated. The formation of hydroxy-aluminosilicates might be expected to influence the bioavailability of aluminium. This has been investigated.[22]

Atlantic salmon fry were exposed to water at about pH 5 containing various concentrations of aluminium and silicic acid as shown below.

	Aluminium total in	Silicon micromoles/litre
TANK 1 (control)	0.85	0.66
TANK 2	7.15	93.06
TANK 3	6.70	24.89
TANK 4	6.22	5.46
TANK 5	6.26	0.60

The mortality of the fish in the various tanks is shown in figure 6. In the control water (tank 1) no fish died in the 96 hour duration of the experiment and this is to be compared to the mortality observed in tank 5 which contained the minimum level of silicic acid with a toxic concentration of aluminium. No deaths were observed in tank 2 in which a toxic level of aluminium was combined with the highest silicon concentration used-93 micromole/l.

The concentration of aluminium in the fish from the various tanks was found by digesting whole fish in nitric acid after external washing. The results are shown below.

It is clear that, in tank 2, silicon at the concentration of 93 micromoles/l prevented the adsorption of aluminium onto gill surface structures and also inhibited systemic absorption. The gills of fish in tanks 1 and 2 were normal;

	Aluminium in fish (micromole/g dry mass)
TANK 1	0.44
TANK 2	0.40
TANK 3	2.04
TANK 4	2.49
TANK 5	2.38

those of fish in the other tanks in which high mortality occurred showed excessive mucus production, sloughing and cellular damage.

It is notable that the silicon concentration at which this inhibition occurred corresponds to the silicic acid concentration required for the formation of imogolite-precursor species – see figure 5.

It seems likely, as calculations suggest, that in the micro-environment proximate to the gill surface, the pH is more alkaline than that of the bulk water and that aluminium is prevented from binding to ligands on the gill surface by stronger binding to silicic acid when hydroxy-aluminosilicates are formed. It is clear that in this experiment the role of silicon as silicic acid was to exclude aluminium from a living creature at the interface between creature and the external environment. Important questions are raised such as: Is this a general role for silicon, not only at the fish gill but also in the mammalian gut? What is the role of the silicic acid normally present in body fluids (5 to 20 micromoles/1) on the toxicity of aluminium that enters the system?

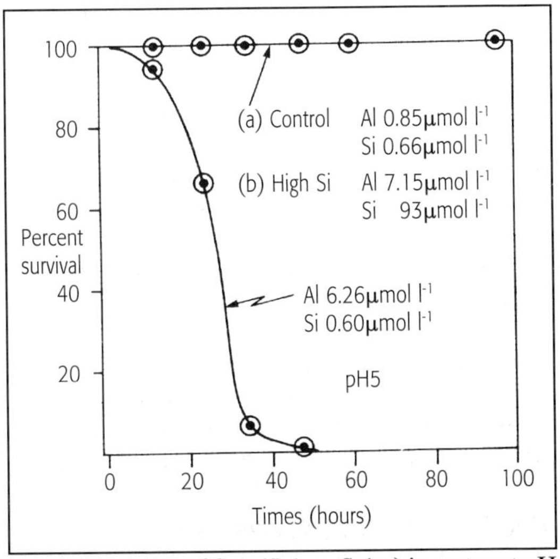

Fig. 6. Mortality of fish (Salmo Salar) in water at pH 5 containing a toxic level of aluminium (ca 7 μMl^{-1}) and with high (93 μMl^{-1}) and low (0.6 μMl^{-1}) silicic acid comparison with control water.

Whilst it is established that, in some species, silicon deficiency produces profound pathological changes to bone and connective tissue and a pronounced growth retardation, no interactions between silicic acid and bio-molecules have been discovered to explain such effects. However, it is becoming increasingly found that available aluminium has powerful effects on a number of important metabolic processes and it seems likely that silicon deficiency exposes biological systems to aluminium. This is certainly the case in some plant species in which the uptake of aluminium (and excessive manganese) in acid soils is reduced by silicic acid.[23]

The Diatom and Silicon Essentiality

One organism for which silicon is considered essential is the diatom. Silicon is used to contruct the hydrated silica frustule but it is also thought to have a role in major metabolic processes such as DNA and chlorophyll synthesis, for on silicon depletion, such synthesis stops and returns when silicon (as silicic acid) is replenished.[24] There have been several suggested biochemical mechanisms to explain such effects such as the involvement of silicic acid in the citric acid cycle between acetyl-CoA and 2-oxoglutarate; in the synthesis of special proteins localised in cell organelles; in chryoslaminaran utilisation and in chlorophyll synthesis. It is difficult to see how silicic acid could be directly involved in any of these metabolic processes and its interaction with an inhibiting metal (aluminium) seems a more likely explanation for some of the observed effects of silicon deficiency. When aluminium, silicic acid and phosphate are present together in solution, there is a change in the binding of alu-

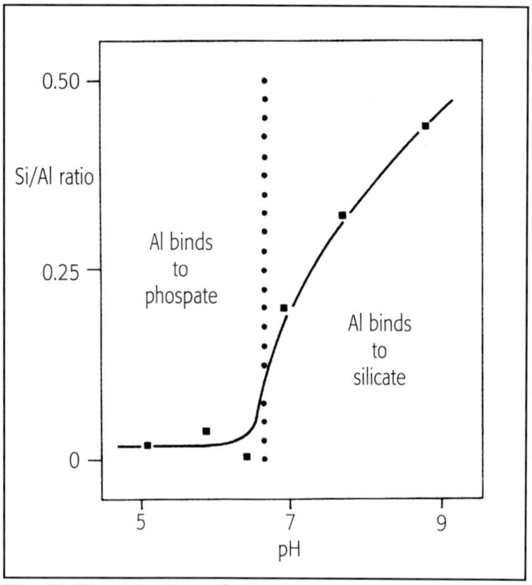

Fig. 7. The binding of aluminium to phosphate and silicic acid as a function of pH. Solutions containing 0.1 mMl^{-1} $AlCl_3$ and 0.5 mMl^{-1} $Si(OH)_4$ passed through a column of aminophosphate resin.

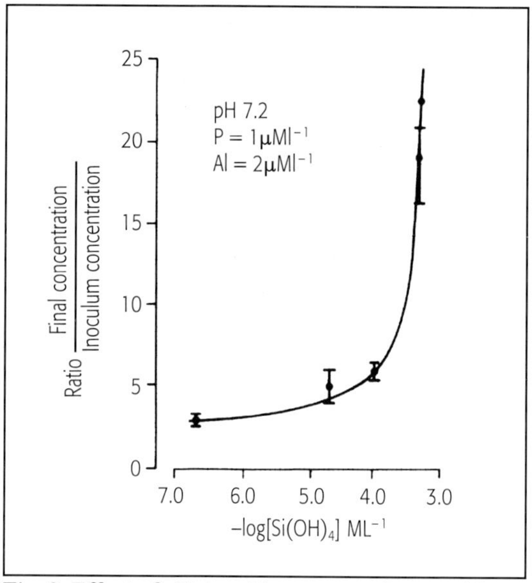

Fig. 8. Effect of silicic acid concentration on chlorophyll a synthesis by Navicula Pelliculosa in a phosphorus-limited culture. Note increase at 10^{-4} silicic acid.

minium from phosphate at below pH 6.6 to silicic acid above pH 6.6 Solutions containing 0.1 mM/l of aluminium and 0.5 mM/l each of silicic acid and phosphate give precipitates in which the Si/Al ratio varied with pH. At pH 7.4 the Si/Al ratio was 0.44 whereas at pH 6.4 the silicon content was negligible and the ratio was 0.02. Experiments in which solutions containing 0.1 mM/l aluminium and 0.5 mM/l silicic acid were passed over a resin having an aminophosphonate functionality confirmed this variability, a switch in the binding of aluminium from phosphate to silicic acid being indicated at about pH > 6.6 – see figure 7. Such a switch may be responsible for some of the symptoms of silicon deficiency in diatoms. The presence of aluminium in culture media has been ignored but in conditions of phosphate limitation, aluminium could reduce the bioavailability of phosphate if silicic acid concentration is low. Experiments with the freshwater diatom, Navicula Pelliculosa, have indicated this to be the case.[25] Figure 8 shows chlorophyll synthesis as a function of silicon concentration in a Navicula Pelliculosa culture at pH 7.2 containing 2 micromole/l aluminium and 1 micromole/l phosphorus. What is especially interesting is that the increase in chlorophyll synthesis is most marked at silicon concentrations above 100 micromole/l – the concentrationn required for stable hydroxy-aluminosilicate formation. In this instance an important role for silicic acid is to inhibit the binding of phosphate by aluminium, maintaining its bioavailability.

Thus the evidence is mounting that an important role for silicic acid is in mediating the biological availability and toxicity of aluminium: indeed this may be its only (but vital) function.

What are the consequences of this intimate relationship between the two elements in the present debate concerning human disease and environmental aluminium?

The Toxicity of Aluminium

Most of our knowledge of the toxic effects of aluminium have been gained as a result of studies of the disorders experienced by dialysis patients when the dialysis fluid contains aluminium greater than about 0.5 micromole/l. The plasma concentration can then rise above the normal level of < 1 micromole/l to, say, 5 micromole/l. In such patients, aluminium accumulates in various tissues including kidney, liver, bone, brain and heart[26, 27] giving rise to pathological conditions such as dialysis encephalopathy which can proceed to dementia and death; dialysis osteomalacic osteodystrophy in which there is defective mineralisation of pre-bone collagen accompanied by bone pain and spontaneous fracture; there is also a microcytic anaemia not responding to iron therapy. (For a review see reference 27). Aluminium also accumulates in the heart and this may be a contributory factor in cardiac hypertrophy often observed in haemodialysis patients.[28, 29] It is generally agreed that aluminium in plasma is carried in the iron transport protein transferrin which will also bind aluminium, although with a significantly lower binding strength than is the case with trivalent iron (log K = 12 for Al and 22 for ferric iron). It is then destined for cells and tissue with high transferrin receptor density.[30]

The major question is what is the mechanism (or mechanisms) of cellular toxicity once aluminium has gained entry to a cell via the transferrin pathway?

It is now clear that an important mechanism is the competition of aluminium with magnesium for binding at key sites at which magnesium is an essential co-factor. Aluminium is known to affect the activity of kinases, polymerases, GTP-binding proteins and other systems which require magnesium. This is due to the binding of aluminium being orders of magnitude stronger than that of magnesium (e.g. it binds about 10^7 times more strongly to ATP than does magnesium) so that nM amounts of aluminium can compete with mM amounts of magnesium. The rate of ligand exchange is often important in biochemical manipulations in which magnesium and calcium are essential co-factor metals and once aluminium is bound, the rate of ligand exchange is 10^4 slower than with magnesium and 10^7 times slower than with calcium.

The situation to be considered is illustrated in figure 9 with aluminium passing from one ligand to another depending on relative binding strength, ligand concentration, pH, competition with other metals etc. – e.g. from citrate (0.1 mM in plasma) to transferrin thence to some intracellular site and eventually to an insoluble depot in bone, bound to phosphate groups on a phosphorylated polymer (thus cross-linked and «insoluble»), or co-deposited with silicic acid (as in senile plaque cores)[31] or phosphate.[32] It may be extracted from some of these depots by the strong chelator desferrioxamine. In the course of this migration, binding at a key site may cause profound disturbance to some biochemical manipulation. One system identified as being particularly vulnerable is the intra-cellular calcium signalling system.[33] Experiments have confirmed that this ubiquitous system is indeed disturbed. The effect of intracellular aluminium on calcium signalling in single mouse pancreatic acinar cells was investigated by measurement of the calcium-dependent chloride ion current using the patch-clamp, whole-cell recording configuration.[33] Intracellular infusion of an aluminium chloride solution was achieved via a fine tube inserted into the patch-clamp pipette. Before the infusion was started, acetylcholine or caffeine were applied externally and evoked pulsatile calcium-dependent chloride currents. After starting infusion with 1 mM $AlCl_3$ solution, external ACh no longer evoked a response, even at 500 nM- tenfold the concentration that evoked a response in the non-infused cells (See Fig. 10). There are many steps from the receptor activation by ACh to the cytoplasmic calcium signal and aluminium could interfere with many of these but, in the work reported, aluminium also inhibited caffeine-evoked intracellular calcium release and it was suggested that the mechanism involved Al inhibiting the caffeine-sensitive, calcium-induced calcium release channel. (See Fig. 11). The fact that aluminium influences the ubiquitous calcium second messenger system may be its prime role in prompting cellular damage although many other systems may also be affected. The single most important point in considering the mechanism of toxicity is the ability of aluminium to displace magnesium at

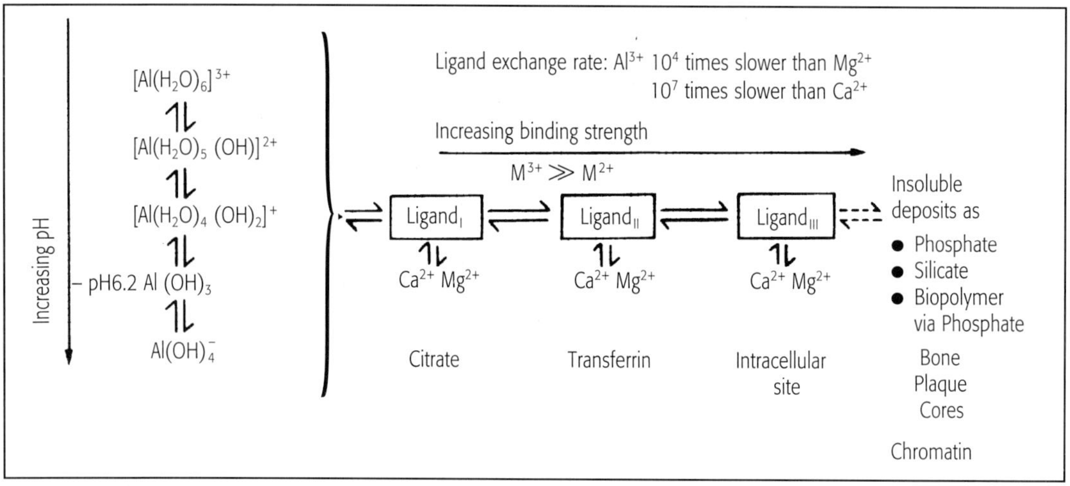

Fig. 9. Aluminium in biology. The key problem – the process of transport and binding.

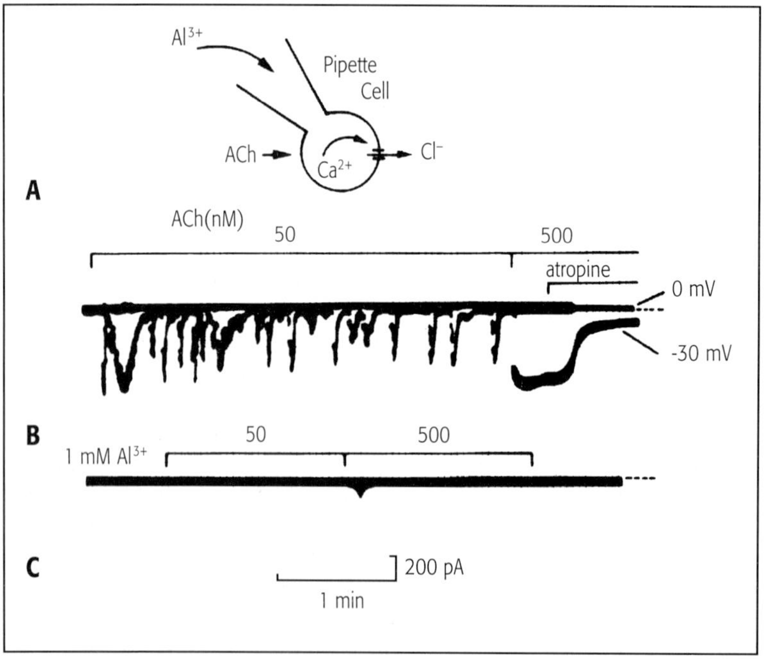

Fig. 10. The effect of acetylcholine (ACh) on Ca^{2+} – dependent Cl^- current in a single internally perfused rat pancreatic acinar cell in the absence and presence of aluminium (1 mMl^{-1} $AlCl_3$) in the pipette solution. A) Shows experimental arrangement. B) Control response to 50 and 500 nM ACh. Atropine (0.1 μM) abolished the effect of 500 nM ACh. C) Shows that in the presence of aluminium, ACh failed to evoke any response.

key binding sites and the fact that, such is its relative binding strength, that Al can so compete in the presence of a large excess of magnesium.

It is not the purpose of this paper to discuss in detail the mechanisms of cellular toxicity of aluminium for other authors in this series will deal with this to greater effect. The questions for attention concern the consequences of the interaction between silicic acid and aluminium on the availability and absorption of aluminium in and from the environment, on its activity *in vivo* and of the consequences of all of this on the etiology and epidemiology of disease states thought to be associated with aluminium.

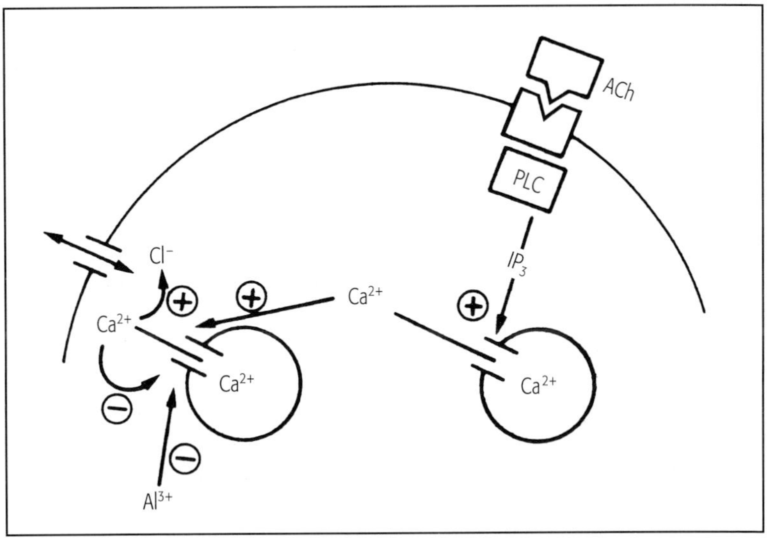

Fig. 11. The mechanism of Ca^{2+} – dependent Cl^- current by aluminium. Aluminium inhibits operation of caffeine-sensitive Ca^{2+} induced Ca^{2+} release channel.

The Consequences of the Al/Si Balance in the Environment External and Internal to Biology

External

The finding that aluminium was excluded from fish in the presence of silicic acid, the knowledge that this exclusion is due to the formation of hydroxy-aluminosilicates, especially at Si concentrations greater than 100 micromoles/l, raises the important question as to the generality of this effect. The concept fits several hitherto unexplained observations, in particular the reduced toxicity of aluminium to plants in the presence of silicic acid and the reduced levels of manganese in plants when silicic acid is present.[35] More recent work by Carlisle on the effect of dietary Si levels on the level of aluminium in the brains of rats on an aluminium supplemented diet showed a high Si diet reducing aluminium uptake with a reciprocal effect of Al on silicon uptake.[36] This is what would be expected given that the exclusion mechanism is the formation of hydroxy-aluminosilicates.

It has been shown in soil chemistry that the concentration of «free» aluminium in interstitial soil water is limited by silicic acid above a certain concentration with the ultimate formation of a solid phase having an Si/Al ratio of 0.5.[37] In ocean chemistry it has been shown that the concentration of aluminium is similarly limited by silicic acid concentration.[38] These findings were not considered in terms of the biological availability of aluminium and, indeed, the sea water experiments over- estimated the solubility of aluminium in the presence of silicic acid because the filtration techniques used would allow colloidal hydroxy-aluminosilicates to pass through and be recorded as «soluble» aluminium. It is now reasonably certain that the colloidal and sub-colloidal hydroxy-aluminosilicate species in which the Si/Al ratio approximate to 0.5 and which form from $pH < 5$ when the silicic acid concentration is > 100 micromoles/l are not readily available to biology.

Some Considerations

With the finding of aluminium, co-localised with silicon, in Alzheimer brains and the conjectured role of aluminium in the disease, it has become of interest to investigate the role of environmental aluminium in the etiology of A.D. and special attention has been given to drinking water. Aluminium salts are used to clarify water from upland areas.

There have been several studies to discover

any relationship that might exist between the aluminium content of drinking water and the incidence of Alzheimer's disease and one of these will be considered here in order to highlight some of the problems raised by such studies. Martyn et al.[39] found an increasing incidence of probable A.D. in areas having aluminium in the drinking water but two criticisms were directed at the study. First, there was no dose response with near-maximum frequency of disease at low aluminium concentration. Second, even at the maximum level of aluminium considered, the aluminium daily ingested from water would, at less than 0.5 mg, be a fraction that ingested from food – 5 to 20 mg. Martyn et al. suggested that the aluminium in water is particularly available. Most foods and drinks contain small concentrations of aluminium and Schenk[40] has provided a list including soft drinks at < 100 – > 2000 micrograms/l; alcoholic drinks at < 100 – > 3000 micrograms/l and food products at < 50 – > 90, 000 micrograms/kg. The total aluminium content may not reflect the amount available for absorption because this will depend on other dietary constituents such as citrate which is known to increase absorption.[41]

The present work suggests strongly that the findings of Martyn and others reveals the effect of silicon in water on the availability and absorption of dietary aluminium. Because of geochemical considerations, water containing significant aluminium contains little silicon and vice versa (see later). It seems highly probable that the preparation of foods and drinks with water containing > 100 micrograms/l Si (as silicic acid) or the co-consumption of such water will limit the gastro-intestinal absorption of aluminium. Solid foods contain silicon with cereals having the most significant content.[42] Oats, for example, contain > 4000 mg Si/kg dry matter. The intake of silicon in the British diet has been estimated at 30 mg/day,[42] with 60% of this being derived from cereals and 20% from water. However, estimates of dietary silicon vary widely from 10 mg/day to as much as 1200 mg/day.[43] Little is known of the availability of the silicon associated with cereals, at least some of which is likely to be derived from phytoliths and hence polymeric. This solid material may adsorb aluminium, reducing availability. The silicon in water is present as ortho-silicic acid and will thus react readily with dissolved aluminium species to form hydroxy-aluminosilicates when the concentration is > 100 micromoles/l Si.[44] If it is assumed that the dietary ingestion of aluminium in food is broadly constant throughout the population, then it is possible that the proportion of this potentially available for absorption will depend significantly on the ingestion of «reactive» silicic acid derived from water. This should be reflected in epidemiological studies. This concept may throw fresh light on an unsolved puzzle – the regional variation in cardiovascular disease. This is discussed in some detail below.

The «Water Factor» in the Regional Variation in Cardiovascular Disease

A marked regional variation in the incidence of cardiovascular mortality in the British Isles has been noted for over a century[45] and subsequently confirmed in several studies.[46] Similar geographic variations have been noted in several other countries,[47] a finding reducing the significance of socio-economic factors and pointing to an environmental agent. These geographic variations are manifested in a difference in cardiovascular mortality rates between hard and soft water areas with the rates in soft water areas being higher than in hard water areas even after adjustment for other factors. Many studies have demonstrated a «water factor» and this has prompted the comment that two possibilities require exploration: «some substance in hard water protects against the disease or that some quality in soft water enhances the disease» [Schroeder in Ref 48]. Soft water is relatively free from dissolved minerals whereas hard water contains many anions and cations other than the major components (calcium and magnesium) primarily responsible for hardness. A major problem in attempting to identify a component of hard water as being a protective agent is that with no ingredient is the human ingestion from water more than a fraction of that ingested from food. This point was made above in respect of aluminium. A second puzzle is that the «protective» effect of hard water is not continuous with degree of hardness (i.e. Ca and Mg content) but reaches a peak at medium hardness – 1.7 mMol/l.

An important clue as to the mechanism prompting the noted higher cardiovascular mortality rate in soft water areas has been provided. A pathological examination of male accident victims and men who had died suddenly of ischaemic heart disease in Glasgow (soft water) and in London (hard water) suggested a factor affecting the myocardium rather than arterial disease with sudden death due to fibrillation and disturbed electrophysiology being more frequent in soft water areas.[49] The «soft water myocardium» may be particularly vulnerable to minor degrees of ischaemia.

An association has been noted between aluminium accumulation in the myocardium and cardiomyopathy with a high prevalence of arythmia and sudden death in haemodialysis patients.[50, 51, 52] In uremic patients, he heart accumulates aluminium [52] and elderly hearts appear to contain more aluminium than the hearts of young accident victims – 0.72 microgram/g dry weight as against 0.3 microgram/g dry weight.[53] In a study of the effects of aluminium on the electrical and mechanical properties of frog atrial muscle, the force of contraction of frog atrial trabeculae was altered by aluminium and the effects on membrane ionic currents and other complex changes suggested that aluminium alters the release of calcium from intracellular, caffeine-sensitive stores.[54]

Note that this is the mechanism suggested for the effect observed of aluminium on calcium mobilisation in mouse pancreatic acinar cells.[34] The intravenous administration of an aluminium acetylacetonate complex to rabbits produced aluminium accumulation in the myocardium (3-4 times control) and a profound cardiomyopathy with multiple areas of interstitial hyperplasia, muscle cell necrosis and myocarditis. There was no evidence of cell alterations in skeletal muscles.[55]

Thus evidence is growing that the heart can be a target organ for aluminium accumulation and that its presence in the heart muscle can promote changes in electrical and mechanical properties. The significance of this to the resolution of the identity of the «water factor» is of great interest.

Silicic acid is a component of hard water, the correlation of its presence with water hardness being given as $r = +0.75$.[56] Studies have indicated a negative correlation between Si concentration in water and cardiovascular mortality ($r = -0.66$) [56] but no mechanistic explanation for such an observation has been possible. However, if silicic acid in water is responsible for inhibiting the absorption of the aluminium in diet and absorbed aluminium accumulates in the heart, producing changes in electro-mechanical behaviour, then geographic variations of the kind observed are to be anticipated. Elford et al.,[46] in reporting a recent study of the geographic variation in heart disease mortality in the UK, ask the question: «Why does a line from the Severn Estuary to the Wash divide the country into one part with high cardiovascular mortality (north of the line) and one with low mortality?». Broadly, this line divides the country between north-western, upland colder, wetter regions of old, well-weathered geology (a catchment geology giving soft water) and young, still-weathering geology in the south-west giving hard water. The table below gives the silicon content of tap water from various locations north and south of this line. In this table, the *Si* concentrations are taken from reference 57 and the I.H.D. risk data from reference 46. The chemistry of the formation of hydroxy-aluminosilicates suggests that aluminium exclusion will be significant only in areas in which the silicic acid concentrations is $>$ 100 micromoles/l. This could account for the observed lack of additional protection afforded by hard water with $>$ 1.7 mg/l Ca and Mg hardness – silicic acid levels can reach maximum concentration in water of medium hardness. The Severn-Wash line dilineates areas with water Si concentrations above and below the critical level of 100 micromoles/l and, broadly, the risk of a major I.H.D. event is seen to be high in low Si, soft water areas. A slow accumulation of aluminium in the heart may then be responsible for the particular electrophysiological vulnerability of the heart in

Table I. **Silicon in tap water in various areas of the U.K. and I.H.D. risk. (Si concentration from reference 57. I.H.D. risk from reference 46).**

Location	Si μmol l^{-1}	Risk of major I.H.D. event per 1000 men/year
Scotland		
Aberdeen	61	
Dundee	68	
Edinburgh	73	
Glasgow (Loch Katrine)	10	9.9
Irvine	7	
Isle of Aran	50	
Isle of Lewis	10	
Renfrewshire	14	
England & Wales North of Line Severn Estuary - Wash		
Bristol	28	
Caerphilly	18	North England 8.4
Cheshire	18	
Knutsford	36	Midlands & Wales 8.0
Leicester	89	
Llandudno	21	
Manchester	25	
Newcastle	28	
Preston	50	
Rugby	7	
Wolverhampton	57	
England South of Line Severn Estuary - Wash		
Cambridge	242	
Didcot	256	4.7
London	147	
Norwich	214	

soft water areas postulated by Crawford.[49] Clearly, as more becomes known of the cellular mechanisms of toxicity of aluminium, its interactions with silicon (silicic acid) and the modified behaviour the or exclusion that appear to result from such interactions become increasingly significant.

The Effects of Internal Silicic Acid

So far, the effect of silicic acid in excluding aluminium from organisms has been considered. But what effect does internal silicic acid have on the cellular toxicity of aluminium that has gained entry? Plasma Si levels appear normally to be in the region of 5 to 20 micromole/l and only in rare cases (such as in dialysis using Si-rich water) does the plasma level rise to the 100 micromoles/l required for the formation of hydroxy-aluminosilicate species. Whether such a level of silicic acid in plasma would moderate the toxic effects of aluminium is a matter still to be decided by appropriate experiment and perhaps by searching for any difference in response of dialysis patients on oral aluminium hydroxide therapy in areas of high and low water silicon.

In dialysis osteomalacia, aluminium is

found at the cement line in affected bone and there is the finding by Carlisle[7] of silicon at the mineralisation front in bone. However, bone does not appear to have been examined for the coincident location of both aluminium and silicon. Such experiments are necessary in order to discover if the association of silicic acid with aluminium *in vivo* eliminates the toxicity of aluminium.

There is evidence from *in vitro* experiments to suggest that elevated plasma silicon levels ($>$ 100 micromoles/l) would influence the toxicity of aluminium. Very preliminary work has been done to repeat the patch-clamp experiments on mouse pancreatic acinar cells which showed that intracellular aluminium inhibited acetylcholine and caffeine-evoked calcium mobilisation.[34] The microinjection of silicic acid itself (100 micromoles/l solution in the pipette) stimulated the cellular response (chloride current) to external ACh, encouraging the formation of well-defined wave patterns. Silicic acid (100 micromoles/l) also completely protected against the inhibitory effect induced by 10 micromoles/l aluminium solution.[58] The stimulatory effect of silicic acid itself observed in these experiments may be due to the «neutralisation» of traces of aluminium present in the system.

A recent patent application entitled «Modulation of Adenylate Cyclase Activity» (International Publication Number WO 90/ 07574, 12 July 1990 to the Ethyl Corporation) claims that silicic acid modulates adenylate cyclase activity in a variety of tissues and that this suggests that silicic acid can be used to modulate hormone or neurotransmitter response in human and veterinary medicine. It is claimed also that the results suggest the use of silicic acid to treat «calcium-related» bone disease. The tissues investigated were liver, kidney, thyroid and bone. In bone tissue, for example, silicic acid was claimed to increase adenylate cyclase activity from 1.0 to 1.8 pMole cyclic AMP per tube at 2.0 mM silicic acid (saturation) and all doses of silicic acid enhanced the effect of PTH on adenylate cyclase activity. The mechanism of the effect was unclear to the investigator who debates that silicic acid may act by binding calcium or magnesium, a proposal considered supported by the fact that silicic acid showed little effect in the presence of ethylene glycol-O, O′-bis(2-aminoethyl)- N,N,N′,N′ tetraacetic acid (EGTA). However, silicic acid will *not* bind calcium or magnesium at physiological pH and a more likely explanation of the effects observed is suppression of the inhibitory effect of traces of aluminium introduced in the tissue preparation process.

There is a danger in interpreting any effects of silicic acid observed *in vitro* experiments in terms of an activity of silicic acid itself, ignoring the strong likelihood of aluminium contamination, the inhibitory effect of aluminium on the system and the removal of that inhibition by silicic acid.

Concluding Remarks

There is a long history of interest in the possible role of silicon in biology and even Pasteur remarked that the element would be found to have an important role. That silicon can be shown to have strong effects *in vitro* and *in vivo* is without doubt but there has been no convincing evidence that these effects are due to some intrinsic essentiality of the element via its direct involvement in metabolic processes. Indeed, the aqueous solution chemistry of silicic acid – the form in which silicon exists in biology – does not suggest any chemistry other than interactions with cations and even these are restricted at physiological pH to aluminium and possibly manganese. A totally new insight into the role of silicon in biology has been provided by the growing awareness of the effect of aluminium on vital biochemical processes and the recognition of the unique interactions between silicic acid and aqueous aluminium species. The formation of hydroxy-aluminosilicates appears to reduce the biological availability of aluminium, excluding the metal from organisms and may also detoxify the metal within organisms under some circumstances. Given the ubiquity and biological redundancy of aluminium, silicon as a natural antagonist, is indeed an essential element.

References

1. Birchall J.D.: *Silicon in Biology*. In (Ed. R.J.P. Williams, J.R.R.F. Da Silva) *New Trends in Bio-Inorganic Chemistry*, Academic Press, 1978.
2. Werner D.: *Silicate Metabolism*. In (Ed. D. Werner), *The Biology of Diatoms*, Blackwell, 1977.
3. Eppley R.W.: The Growth and Culture of Diatoms in Ref. 2.
4. Okuda A., Takahashi E.: in *The Mineral Nutrition of the Rice Plant*, Symp. Intern. Rice Res, 123-146, John Hopkins Press, Baltimore, 1964.
5. Schwarz K., Milne D.B.: *Nature*, **239**, 333-334, 1972.
6. Carlisle E.M.: *Science*, **178**, 54-156, 1972.
7. Carlisle E.M.: *Science*, **167**, 179-180, 1970.
8. Schwarz K.: *Proc. Nat. Acad. Sci.*, USA, **70**, 1608-1612, 1978.
9. Schwarz K.: in (Ed. G. Bendz, I. Lindqvist) *Biochemistry of Silicon and Related Problems*. Plenum Press, New York, 1978.
10. Loeper J. et al.: *Giornale di Clinica Med.*, **47**, 596-605, 1967.
11. Silicon Biochemistry, Ciba Foundation Symposium 121, (Ed. D. Evered, M. O'Connor), Wiley, 1986.
12. Carlisle E.M., Alpenfels W.F.: *Fed. Proc.*, **43**, 680, 1984.
13. Viola R.E., Morrison R.E., Cleland W.W.: *Biochemistry.*, **19**, 3131-3157, 1980.
14. Iler R.K.: *The Cemistry of Silica*, Wiley, 1979.
15. Farmer V.C., Frazer A.R., Tait M.: *Geochim. Cosmochim. Acta*, **338**, 146-148, 1979.
16. Chapppell J.S., Birchall J.D.: *Inorg. Chim. Acta*, **153**, 1-4, 1988.
17. Birchall J.D., Thomas N.L.: in (Ed. F. Glasser), *The Chemistry and Chemically-Related Problems of Cement*, 305-315, Brit. Ceram. Proc. No. 35, Brit. Ceram. Soc., 1984.
18. Driscoll C.: *Environ. Health Perspec.*, **66**, 93-104, 1985.
19. Witters H.E.: *Aqua Tox.*, **8**, 197-210, 1986.
20. Muniz I.P., Leivesad H.: Proc. Int. Conf., Impact. Acid Precip. (Norway, SNSF Project), 1980.
21. Driscoll C.T., Baker J.P., Bisogni J.J., Schofield C.: *Nature.*, **264**, 161-164, 1980.
22. Birchall J.D., Exley C., Chappell J.S., Phillips M.J.: *Nature.*, **338**, 146-148, 1989.
23. Peaslee D.E., Frink C.R.: *Soil. Sci. Soc. Am. Proc.*, **33**, 569-571, 1969.
24. See Reference 2.
25. Exley C., Birchall J.D.: Submitted, 1991.
26. Roth A., Nogues C., Galle P., Drueke T.: *Virchows Arch. (Pathol. Anat.)*, **405**, 131-140, 1984.
27. Kerr D.N.S., Ward M.K.: in (Ed. H. Sigel, A. Sigel), *Metal Ions in Biological Systems*, Vol. 24, *Aluminium and its Role in Biology*, 217-258, Marcel Dekker Inc., 1988.
28. London G.M. et al.: *Amer. J. Kidney Disease*, Vol. XIII, 75-83, 1989.
29. Levine S.N., Sonnier G.B., Abreo K.: *Toxicology*, **65**, 137-148, 1990.
30. Candy J.M., Edwardson J.A., Faircloth R. et al.: *J. Physiol.*, **34**, 391, 1987.
31. Candy J.M., Oakley A.E., Klinowsky J. et al.: *Lancet*, **i**, 354-357, 1986.
32. See Reference 26.
33. Birchall J.D., Chappell J.S.: *Lancet*, **i**, 1008-1010, 1988.
34. Wakui M., Itaya K., Birchall J.D., Petersen O.H.: *FEBS*, **267**, 301-304, 1990.
35. See Reference 23.
36. Carlisle E.M., Curran M.J.: *Alzheimer Disease and Associated Disorders*, **1** (2), 83-89, 1987.
37. Farmer V.C. in Reference 11, 4-23.
38. Wiley J.D.: *Marine Chemistry*, **3**, 227-240, 1975.
39. Martyn C., Baker L.J.P., Osmond C. et al.: *Lancet*, **i**, 59-62, 1989.
40. Schenk R.U. et al.: *Composition and Consequences of Aluminium in Water, Beverages and other Ingestibles*, in (Ed. T. Evans) *Environmental Chemistry and Toxicity of Aluminium*, 247-269, Lewis Publishers, Michigan, USA, 1989.
41. See Janet L. Gregor in Reference 27 (199-215).
42. Bowen H.J.M., Peggs A.: *J. Sci. Food. Agric.*, **35**, 1225-1229, 1984.
43. Hamilton E.L., Minski M.J.: *Sci. Total Environ.*, **1**, 375-394, 1973.
44. Birchall J.D., Chappell J.S.: *Lancet*, **i**, 953, 1989.
45. Haviland A.: *The Geographic Distribution of Heart Disease and Dropsy in England and Wales*, Churchill, London, 1871.
46. Elford J., Thomson A.G., Phillips A.N., Shaper A.G.: *Lancet*, **i**, 343-346, 1989.
47. Masironi R., Pisa Z., Clayton D.: Myocardial Infarction and Water Hardness in the W H O Myocardial Infarction Registry Network. *Bulletin of the World Health Organisation*, **57** (2), 291-299, 1979.
48. Schroeder H.A.: *J. Amer. Med. Assoc.*, **195**, 81-85, 1966.
49. Crawford T., Crawford M.D.: *Lancet*, **i**, 229-232, 1967.
50. See Reference 26.
51. Webb J., Kirk K.A., Niedermeier W. et al.: *J. Mol. and Cellular Cardiology*, **6**, 383-394, 1974.
52. See Reference 28.
53. Lyon T.D.B., Fell G.S.: Trace Element Unit, Institute of Biochemistry. Royal Infirmary, Glasgow, Personal Communication, 1991.
54. Meiri H., Shimoni Y.: *J. Pharmacol.*, **102**, 483-491, 1991.
55. Zatta P., Giordano R., Corain B., Favarato M., Bombi G.G.: *Toxicology Letters.*, **39**, 185-188, 1987.
56. Pocock S.J., Shaper A.G., Cook D.G., Packham R.F., Lacy R.F., Powell P., Russel P.F.:

British Regional Heart Study: Geographic variations in Cardiovascular Mortality and the Role of Water Quality. *Brit. Med. Journal.*, **280**, 1243-1249, 1980.

57. Dobie J.W.: *Scot. Med. J.*, **27**, 10-16, 1982.
58. Petersen O.H., Petersen C.: Dept. of Physiology, University of Liverpool. Personal Communication, 1991.

Section III

MEDICINE

Aluminum Intoxication Recognition and Treatment

A.C. Alfrey

Although aluminum is ubiquitous, being the fifth most common element in the earth's crust, normally humans, other animals, trees and fish have had limited exposure to this element because of the insoluble nature of most aluminum compounds. In areas of acid soil where the solubility of aluminum compounds is markedly enhanced plants have adapted means of limiting their exposure to soluble aluminum compounds. Toxicity is averted by some plants by the formation of mucilage by the plant roots, which chelates the aluminum preventing uptake by the roots.[1] Other plants, such as the ferns and the tea plant, are aluminum accumulators in that they chelate or immobilize any aluminum taken up through their roots in their leaves as aluminum oxides.[2] However, animals are much less adapted to the prevention of uptake and toxicity from soluble aluminum compounds. In acidified lakes with high aluminum levels aluminum has been shown to exert toxicity on fish by interfering with gill function in regards to transfer of water and oxygen.[3] High aluminum levels in water have also been suggested to be toxic to birds by causing defects in shell formation and impairing breeding of certain sea birds.[4]

Normal Aluminum Physiology

Aluminum Exposure

Human aluminum intoxication has also largely been prevented by the insoluble nature of aluminum compounds in association with the fact that the lungs, skin and gastrointestinal tract have represented formidable barriers to aluminum absorption. Normally only 2 to 3 mg of aluminum is ingested as a normal contaminate in food sources.[5] However, aluminum is used as a filler in certain foods such as pickles and cheese and is the major constituent in backing powder. In addition aluminum is commonly added to water at municipal water treatment facilities to flocculate and remove turbidity which is present in the water. This can leave aluminum residues as high as 4 mg/liter.[6] Therefore under certain circumstances aluminum ingestion can be considerably greater than just the aluminum normally present as a contaminate in food and water sources. The small amount of aluminum that is normally absorbed from the gastrointestinal tract (5-10 μg/day) is largely eliminated by the kidneys.[7]

Under usual circumstances aluminum which is inhaled by the lung is largely exhaled and the small amount retained is trapped in the pulmonary tissue preventing the systemic transfer of this element. This is supported by the finding that pulmonary aluminum levels are much higher than any other tissue levels, increase with ageing and do not correlate with any other tissue aluminum levels.[8,9] Under unusual circumstances when aluminum levels are extremely high in the inhaled air or aluminum is present as fumes some aluminum absorption through the lungs may occur.[10] Whether enough aluminum uptake occurs to increase the body burden of this element is unknown. The skin appears to be highly impervious to aluminum compounds even when the exposure is great as occurs with the application of numerous antiperspirants.

Thus the total body aluminum in adults is maintained at about 35 mg and does not increase with ageing. The normal biological fluids and tissue aluminum values found in adults are shown in Table I.

Gastrointestinal Absorption of Aluminum

The majority of studies would support that aluminum is absorbed passively by the gastrointestinal tract. There appears to be very little

Table I. **Normal aluminum values in biological fluids and tissues.**

Biological Fluids	Values (μg/liter)
Plasma	2.4 ± 0.4
Cerebrospinal fluid	6.8 ± 4.5
Bile	3.0 ± 1.0
Urine	7.1 ± 1.0*
Biological Tissues	Values (mg/kg dry wt.)
Brain	2.2 ± 1.3
Heart	1.0 ± 0.8
Muscle	1.2 ± 1.2
Liver	4.1 ± 1.7
Spleen	2.6 ± 2.1
Lung	43.0 ± 44.0

Values = mean ± 1 SD * = μg/gram creatinine

if any transcellular absorption but rather aluminum is absorbed solely through the paracellular pathways.[11,12] A major problem studying aluminum absorption is the small fractional absorption of aluminum which occurs through the gastrointestinal tract and lack of a suitable aluminum isotope to estimate absorption. Therefore, indirect means of estimating aluminum absorption, such as changes in urinary and blood aluminum levels, have been used to estimate aluminum absorption under various conditions. As of this time only four factors have been clearly shown to modulate aluminum absorption. These are the amount of aluminum ingested,[7] the solubility of the aluminum compound,[13] the integrity of the tight junctions of the small intestine[11,12,14] and the uremic state.[15] By increasing aluminum intake from mg quantities to gram quantities aluminum absorption can be increased by 20 to 50 times normal to as high as 500 μg/day.[7,16,17] More soluble aluminum compounds, such as aluminum lactate compared to aluminium hydroxide, are also more readily absorbed.[13] The combination which appears to have the greatest influence on aluminum absorption is administering a soluble aluminum compound in association with opening the tight junction between small intestinal cells. This is exemplified by administering aluminum citrate.[11,14] Aluminum citrate is not only relatively soluble as compared to other aluminum compounds but citrate, probably as a result of chelating calcium necessary for the maintenance of tight junction, opens the paracellular pathways. This result in a very rapid and markedly enhanced aluminum absorption. Recent studies would further show that normal individuals chronically ingesting calcium citrate, given as a calcium supplement, also have enhanced aluminum absorption from normal dietary sources.[18] The uremic state, in experimental animals, has also been shown to augment the gastrointestinal absorption of aluminum although the mechanism has not been elucidated. A number of other factors such as vitamin D,[19] parathyroid hormone[20] and the status of the body iron stores[21] have been suggested to effect aluminum absorption although the evidence supporting that these factors have an effect on aluminum absorption is not very convincing.

Elimination of Aluminum

As stated above the kidneys appear to be largely if not solely responsible for the elimination of any absorbed aluminum. Even under conditions of large parenteral loads of aluminum biliary aluminum increases only modestly and accounts for well less thant 1% of the elimination of an aluminum load.[22] Aluminum would appear to be excreted through the kidney by glomerular filtration. However, its clearance is only approximately 5 to 10% of the glomerular filtration rate. This would appear to be a consequence of the fact that the majority of aluminum is present in the blood, bound to transferrin, which would not be filtered. The small amount of aluminum that is filtered is probably complexed to citrate which accounts for the low fractional excretion of this element. The kidney's ability to excrete aluminum is limited and if large parenteral loads of aluminum are administered varying amounts of the administered aluminum is retained.[23] Under usual conditions it would appear that the kidney is able to eliminate any aluminum absorbed through the gastrointestinal tract. However, when aluminum compounds are given orally with citrate, at least in animal studies, the kidney ability to excrete aluminum can be exceeded and some of the aluminum is retained in tissues.[24] This has not yet been documented to occur in humans with normal renal function.

Aluminum Toxicity

Aluminum Loading in Humans

With very rare exceptions total body aluminum excess in humans has only been shown to occur in association with either parenterally administered aluminum compounds and/or compromised renal function. This was first recognized in uremic patients chronically dialyzed with aluminum contaminated dialysate.[25, 26] This resulted in the transfer of aluminum from the dialysate to the blood during the dialysis procedure.[27] As a result of plasma binding of the aluminum the transferred aluminum was retained and the gradient from dialysate to blood maintained further enhancing aluminum transport. The lack of kidney function also prevented the elimination of the administered aluminum from the body during the interdialytic period. It was subsequently shown that body burden of aluminum could also be increased in patients with normal renal function by administering large aluminum loads parenterally.[23] This was initially shown in patients with small bowel disease maintained with chronic total parenteral nutrition. The protein source, a casein hydrolysate, was subsequently shown to be contamined with aluminum. Total body burden of aluminum has also been shown to be increased in both adults and children who had received large amount of intravenously administered albumin.[28, 29] Albumin has been shown to be highly contaminated with aluminum which results from the procedures used to separate it from other serum proteins and store it.[30, 31]

The kidney's ability to eliminate any absorbed or administered aluminum is critical in the prevention of an increased body burden of aluminum. Approximately 85% of non-dialyzed patients with end-stage renal disease, who had no known history of ingesting aluminum containing compounds, were found to have elevated tissue stores of aluminum.[8, 9] Although enhanced absorption of aluminum which normally contaminates food and water may have been partly responsible for this finding,[15] reduction of renal function compromising the ability to eliminate the absorbed aluminum seems to be the most likely cause for the increased tissue stores of aluminum in these patients. The combination of reduction of renal function in association with the orally administration of aluminum compounds can markedly increase the body burden of aluminum.[8, 9] This is even further intensified by the concomitant administration of aluminum compounds and citrate as a result of the citrate markedly enhancing aluminum absorption as discussed above.[11, 12, 14] It has still not been clearly established whether individuals with normal renal function can have an increased body burden of aluminum resulting from orally administration of aluminum compounds. There has been only one patient reported with normal renal function who had increased bone stores of aluminum which appeared to occur as a result of ingesting aluminum containing antacids.[32]

Virtually all tissue stores of aluminum are increased in the state of aluminum excess with bone and liver aluminum levels affected the greatest. Although plasma aluminum levels increase with aluminum loading, they do not correlate well with other tissue stores of aluminum. Plasma aluminum levels would tend to correlate better with active aluminum loading than the established body burden of aluminum.[33] It is very unusual to find plasma aluminum levels greater than 150 μg/liter in uremic patients whose only exposure is orally ingested aluminum compounds. In patients with values greater than 150 μg/liter one should consider that loading is occurring as a consequence of the individual receiving a parenteral load of aluminum through either aluminum contaminated dialysate or intravenous fluids. Another cause of of extremely high plasma aluminum levels, at times in excess of 1000 μg/liter, is the concomitant administration of oral aluminum compounds and citrate.[14, 34] In patients with normal renal function orally administered aluminum compounds rarely cause an elevation of plasma aluminum levels to greater than 10 μg/liter. However, when given parenteral aluminum loads, plasma aluminum levels can increase in excess of 100 μg/liter in individuals with normal renal function.[22]

A useful method of estimating body burden of aluminum is measuring the rise in plasma aluminum levels that occurs following the administration of deferoxamine, an iron and aluminum chelator.[35] Although it is unknown from which body stores aluminum is chelated and responsible for the rise in plasma aluminum levels there is a reasonable correlation between the increase in plasma aluminum levels and bone aluminum stores.[35] It is unknown what if any use deferoxamine administration would be in estimating body aluminum burden in individuals with normal renal function by measuring either change in plasma or urinary aluminum excretion following its administration.

Clinical Features of Aluminum Intoxication

It has been reasonably well established that aluminum can exert toxicity on the neurological,[25] skeletal[36, 37, 38] and hematological[39] systems in humans with advanced renal failure. It is less well established what if any toxicity and under what circumstances aluminum can be toxic in individuals with normal renal function.

Neurological Toxicity

The neurotoxicity for aluminum was initially described in 1886[40] and subsequently rediscovered in 1937.[41] These studies showed that the local administration or application of aluminum to the brain caused the animal to develope seizures. However, renewed interest in aluminum neurotoxicity occurred when it was found that in certain species of animals, cats and rabbits, neurofibrillary tangles developed from both local and systemic exposure to aluminum.[42-47] This resulted in the suggestion that aluminum could be involved in the pathogenesis of Alzheimer's disease.[42] Although some investigators have reported brain aluminum to be modestly increased in the brains of patients dying from Alzheimer's disease most studies do not support a primary role for aluminum in the pathogenesis of Alzheimer's disease.[48]

The first report suggesting that aluminum could cause an encephalopathy in humans was in 1962.[49] The patient described had worked in an aluminum processing plant and developed a fatal dementing neurological disease. Brain aluminum, as well as other tissue stores of aluminum were found to be markedly increased. A second patient with a neurological disease and elevated brain aluminum levels was described in 1975.[50]

In 1972 a new and distinct neurological disease which strongly resembled the neurological syndrome which occurred in the patient described in 1962 was described in dialyzed uremic patients.[51] This syndrome (dialysis encephalopathy or dialysis dementia) was characterized by an intermittent speech disturbance which was a stuttering or stammering that was frequently first noted, and often intensified later in the course of the disease, immediately following dialysis. The speech disturbance was associated with personality changes, parietal lobe findings such as directional disorientation, seizures and auditory and visual hallucinations. As the disease progressed, the patient developed asterixis and myoclonus. Over a 7 to 9 month period following the onset of symptoms, the patient became totally mute and was unable to perform any purposeful movements. Death rapidly ensued. The symptoms and course of this disease were strikingly similar between patients and the disease was soon described in numerous patients from dialysis centers throughout the world.[52-58] It was recognized that this disease occurred in epidemic proportions in some dialysis centers whereas in dialysis centers in other geographical areas it was rarely if ever seen. This suggested the possibility that dialysis encephalopathy was caused by some environmental toxin. In 1976 it was reported that one trace element, aluminum, was consistently higher in brain gray matter in patients dying of dialysis encephalopathy than in dialysis patients dying of other causes and in non-uremic controls.[25] This finding was subsequently confirmed by other groups.[54,56] Additional evidence increminating aluminum in the pathogenesis of dialysis encephalopathy were large epidemiological studies carried out in United Kingdom,[59,60] France[61] and the United States[57] which showed that dialysis centers with a high incidence of dialysis encephalopathy had large amounts of aluminum in the water used to prepare the dialysate. In contrast, centers with a low incidence of dialysis encephalopathy, the water used to prepare the dialysate was either free or very low in aluminum. Further evidence supporting aluminum as the toxin responsible for dialysis encephalopathy was obtained when the number of cases of dialysis encephalopathy profoundly decreased when aluminum was removed from the water used to prepare the dialysate. The final evidence for aluminum neurotoxicity comes from more recent reports showing that aluminum toxicity can be reversed by chelating patients with dialysis encephalopathy.[62,63]

An acute form of aluminum neurotoxicity has recently been described in uremic patients.[34,64] It occurs under three different conditions. These include dialysis performed with dialysate highly contaminated with aluminum, the oral administration of citrate compounds in association with aluminum compounds[34,64] and extreme elevation of plasma aluminum levels in association with deferoxamine therapy.[65] It would appear that virtually all non-dialyzed uremic adults[34,64] and children[66] who have developed aluminum neurotoxicity have done so from the ingestion of the combination of citrate and aluminum compounds. Although the patients may have some mild neurological findings of aluminum neurotoxicity, like the speech disturbance described above, more often there is an acute, explosive onset of symptoms. The major reported findings in adults include agitation, confusion, myoclonic jerks, grand mal seizures, obtundation, coma and death.[34] Aluminum neurotoxicity in children tends to be somewhat different in that it is manifested by regression of verbal and motor skills.[66] This is exemplified by the finding that children who have just started to walk and talk may lose these abilities.

If acute toxicity occurs as a consequence of high aluminum levels in the dialysate it usually develops during the dialysis procedure. Symptoms related to deferoxamine therapy usually occur during or immediately following a dialysis where deferoxamine was administered.[65] Although onset of symptoms of acute aluminum intoxication may occur after months of ingestion of the combination of citrate and aluminum compounds, it is not uncommon for symptoms to develope in a matter of weeks following the combined ingestion of these two compounds.[34]

Diagnosis of aluminum neurointoxication is largely based on a history of either oral or parenteral exposure to aluminum in a uremic patient, demonstration of the classical clinical features and exclusion of other neurological conditions by appropriate studies. The most useful laboratory test supporting the diagnosis of aluminum neurotoxicity is the electroencephalogram.[26] Unlike most metabolic encephalopathy, including uremic encephalopathy, which exhibits a generalized slowing of the EEG, in aluminum intoxication the background rhythm is relatively normal with multifocal bursts of slow or delta waves frequently accompanied by spike activity (Fig. 1). The CT scan may be normal or only show mild cortical

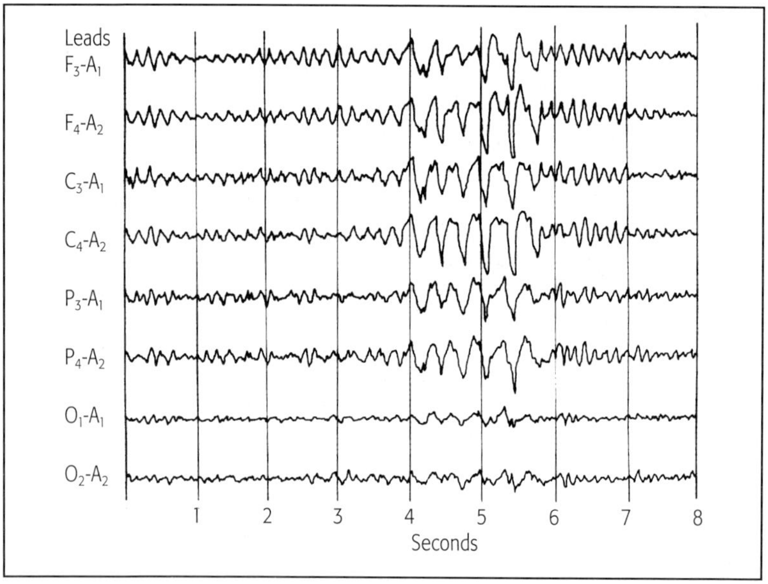

Fig. 1. EEG Alterations Found in Dialysis Encephalopathy. Multifocal bursts of delta and occasional spike activity with relatively normal background rhythm.

atrophy. There are no cerebrospinal fluid abnormalities and aluminum levels in CSF are not increased. Histological changes that occur in the brains of patients dying of dialysis encephalopathy are non-specific and similar to changes seen in dialysis patients dying of other causes.[67] Plasma aluminum levels in chronic intoxication are increased and frequently in the range of 100 to 200 µg/liter. With acute aluminum intoxication plasma aluminum levels are usually in excess of 500 µg/liter.

Skeletal Toxicity

Aluminum skeletal toxicity is characterized by osteomalacia, bone pain, pathological fractures, proximal myopathy and a failure to respond to any form of vitamin D therapy.[35-38] Laboratory studies reveal decreased or absent bone formation rate, relatively low parathyroid hormone levels, usually normal alkaline phosphatase levels and mildly elevated serum calcium levels. Unlike virtually all other osteomalacic states serum phosphorus levels are either normal or more commonly slightly elevated. Evidence for the skeletal toxicity of aluminum is as convincing as it is for aluminum neurotoxicity. The first evidence that aluminum caused bone disease came from European epidemiological studies that showed that fracturing osteomalacia tended to occur in association with dialysis encephalopathy in dialyzed uremic patients in areas with high aluminum levels in the water used to prepare the dialysate.[59, 60] As with encephalopathy, it was shown that fracturing osteomalacia could be largely prevented or would spontaneously heal by removing aluminum from the dialysate.[63, 68] It was subsequently shown that bone aluminum levels were markedly increased and higher than found in other types of renal osteodystrophy.[38] Finally experimentally aluminum was shown to cause osteomalacia in rats,[69] dogs[70] and pigs.[71] Besides parenteral exposure to aluminum, oral aluminum administration can also cause osteomalacia in uremic patients. Two studies in which bone biopsies were randomly done in large dialysis populations showed that 25 to 30% of dialyzed uremic patient having no parenteral exposure to aluminum but routinely receiving oral aluminum containing phosphate binding gels had aluminum-associated bone disease.[72, 73] It has further been shown that virtually all osteomalacia which occurs in uremic patients is secondary to aluminum intoxication and that this form of bone disease does not result from uremia per se.[74]

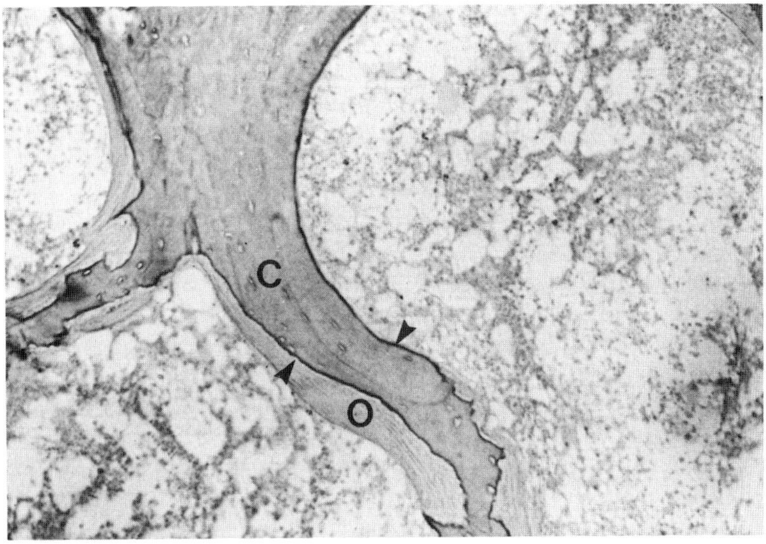

Fig. 2. Bone Histology Present in Aluminum Associated Bone Disease. Classical features of osteomalacia with marked increase in non-calcified osteoid (O) tissue. Aluminum deposition is defined by the dark line (arrows) (Aluminum-stained) between calcified (C) and non- calcified bone (O).

Diagnosis of aluminum associated bone disease is based on the clinical features of bone pain, pathological fractures, usually of the ribs and femoral neck, and proximal myopathy. Laboratory confirmation is made on bone biopsy showing classical osteomalacia with aluminum, as determined by aluminum staining, present between the junction of calcified and non-calcified bone (Fig. 2) and markedly increased quantitated bone aluminum levels.[75] However, in a smaller number of patients an aplastic type of bone disease characterized by absent bone formation and no hyperosteoidosis may be present.[35-38]

Hematopoietic Toxicity

Aluminum has also been shown to be toxic to the hematopoietic system.[39, 76, 77] In uremic patients aluminum has been shown to cause a microcytic, hypochromic anemia in the setting of normal iron stores. This type of anemia usually occurs in the setting of other evidence of aluminum intoxication, such as bone disease or encephalopathy. Aluminum has also been shown to cause an anemia in rats and rabbits.[78, 79] The anemia responds rapidly to aluminum chelation with deferoxamine.[80] In addition aluminum induced anemia is resistant to erythropoietin therapy which can be reversed by deferoxamine.[80]

Metastatic Calcification

Aluminum along with other element such as lead and iron forms colloids in association with phosphate and hydroxides at physiological pH. Because of this characteristic Selye and his group found that injection of aluminum and similar element caused an acute elevation of serum calcium and phosphorus and metastatic calcification in rats when given in the setting of a subcalcifing dose of vitamin D.[81, 82] This phenomena was termed calciphylaxis with vitamin D being the sensitizer and aluminum representing the challenger. Visceral calcification which was common in the early years of dialysis, and which is now infrequent, contained large amounts of aluminum in the deposits. It seems possible that aluminum administered parenterally during dialysis, as a result of aluminum contaminated dialysate, exceeding the ability of transferrin to bind aluminum and formed colloids in association with phosphate and hydroxide and incited a chronic calciphylaxis response. This may have played an integral role in the production of the visceral calcium-phosphate deposits which occurred in these patients.

Aluminum Speciation Role in Toxicity

The speciation of aluminum may also be

important in determining the toxicity of aluminum for the various systems described above. As discussed by Martin[83] aluminum because of its high formation constant with transferrin, citrate, phosphate and hydroxides can not exist in plasma in a free form. Formation constants are greater for aluminum with transferrin and citrate than hydroxides. Therefore, in plasma, aluminum would exist primarily in association with transferrin and secondarily with citrate. In view of the association between acute aluminum neurotoxicity and high levels of small molecular species of aluminum complexed to citrate and deferoxamine it is suggested that these species most readily cross the blood brain barrier and are responsible for neurotoxicity of this element. Aluminum effect on erythropoiesis and its interference with iron utilization has recently been studied *in vitro*.[84] Whereas aluminum alone had no effect on human erythroid cultures, aluminum transferrin mixture caused a dose dependent inhibition of erythroid colony growth. This would be consistent with aluminum-transferrin complexes being responsible for the hematopoietic toxicity of aluminum noted *in vivo*. Some evidence has been presented suggesting that aluminum forms coordinated complexes between carboxyl groups of collagen fibers in association with oxygen, cross linking the collagen fibril and inhibiting bone induction properties.[85] This type of aluminum complex might explain the mechanism of aluminum induced bone disease. Finally as discussed above circulating colloids composed of aluminum, hydroxides and phosphate may have been important in the pathogenesis of metastatic calcification previously commonly seen in dialysis patients.

Management of Aluminum Toxicity

Prevention of Aluminum Intoxication

A major impetus has been directed at the prevention of aluminum intoxication. The only patient populations that have been shown to be at risk of developing aluminum intoxication are patients with advanced renal failure and individuals with normal renal function who have received parenterally administered aluminum compounds over an extended period of time. Parenterally administered aluminum should be avoided when possible. There are now standards for acceptable levels of aluminum in dialysate, < 10 μg/liter, which are easily achievable. For total parenteral nutrition casein hydrolysate has been replaced with aluminum free amino acid solutions for the protein source. The major potential source of parenteral exposure to aluminum is now from albumin solutions. Efforts are being made to modify the manufacturing process to reduce the aluminum contamination.[30, 31]

Currently the major potential source of aluminum intoxication is from orally administered aluminum compounds commonly given to individuals with chronic renal failure to reduce phosphate absorption from the gastrointestinal tract to prevent hyperphosphatemia. Calcium carbonate and calcium acetate are increasingly being used as phosphate binders replacing aluminum binders.[86, 87] Although the calcium salts are not as effective as aluminum in binding phosphate they are frequently adequate for controlling phosphate and even when they are not adequate alone they may reduce the amount of aluminum that also has to be given to control serum phosphorus levels. To avoid toxicity patients with advanced renal failure receiving aluminum compounds should have plasma aluminum levels serial measured and aluminum dosage modified if plasma aluminum levels exceed 40 to 50 μg/ liter.

Treatment of Aluminum Intoxication

Treatment of aluminum intoxication initially involves the removal of all parenteral and oral exposures to aluminum. Following removal of parenteral exposure through the dialysate a number of patients with osteomalacia and far fewer with dialysis encephalopathy have recovered from aluminum intoxication without additional treatment.[63, 68, 88] Similarly patients developing bone disease while on total parenteral nutrition with aluminum contamined solutions have shown improvement following the discontinuation of aluminum exposure.[89] Chelation therapy for aluminum intoxication has only been widely used and adequately evaluated in patients with advanced renal failure usually requiring dialysis.[35, 63, 88, 90] A number of different approaches to chelation have been recommended. Unlike iron, the amount of aluminum chelated appears to be determined by the size of the chelatable aluminum pool and not the dose of deferoxamine administered.[91-93] It has been found that 1 or 2 grams of deferoxamine is equally as effective as 4 grams in chelation of aluminum as determined by the rise in plasma aluminum following chelation. The most effective spacing between dosage of deferoxamine has not been clearly defined so it is unclear whether 1 to 2 grams of deferoxamine should be given once or three times weekly intravenously during the last hour of dialysis. Deferoxamine is equally effective when given intramuscular or intraperitoneal as when given intravenously.[93] Treatment of aluminum associated bone disease with deferoxamine usually results in symptomatic improvement in regards to bone pain and proximal muscle weakness in 2 to 3 months. However, treatment should be continued for around 6 months and discontinued when there is only minimal rise in plasma aluminum levels following deferoxamine administration demonstrating depletion of the chelatable aluminum pool. Symptomatic improvement has been found to occur as rapidly in patients given 0.5 grams as patients given 6 grams weekly dose of deferoxamine.[94] Aluminum induced anemia responds in a similar time period as symptomatic bone disease. In contrast aluminum associated neurological toxicity requires a more extended duration of treatment with deferoxamine. Although progression of the neurological disease may stabilize, improvement frequently requires 10-12 months of continuos treatment with deferoxamine. With adequate treatment EEG abnormalities may resolve making EEG measurement important in determining duration of therapy. Following discontinuation of therapy, especially if too short, recurrence of aluminum neurotoxicity is not uncommon.

Therapy with deferoxamine is not without some risk. As described above patients with marked aluminum overload may develope acute neurological toxicity following deferoxamine therapy. Acute infections with Yersinia and mucormycosis have also been reported in patients receiving deferoxamine.[95, 96] It is felt that feroxamine is an iron supplier for these non-siderophore producing organisms promoting their growth. Therefore, prior to treatment diagnosis of aluminum intoxication should be reasonably well established and evidence of aluminum overload demonstrated either by a rise of plasma aluminum greater than 150 to 200 μg/liter following deferoxamine infusion or increased bone aluminum levels in association with classical histological findings of aluminum associated bone disease.

References

1. Horst W.J., Wagner A., Marchner H.: Mucilage protects root meristems from aluminum injury. *Z. Pflanzenphysiol.*, **105**, 435, 1982.
2. Matsumoto J., Hirasawa E., Morimura S., Takahashi E.: Localization of aluminum in tea leaves. *Plant Cell Physiol.*, **17**, 627, 1976.
3. Driscoll C.T., Bazker J.P., Bisogni J.J., Schofield C.L.: Effect of aluminium on fish in dilute acidified waters. *Nature*, **17**, 627, 1980.
4. Nyholm N.E.I.: Evidence of involvement of aluminum in causation of defective formation of eggshells and of impaired breeding in wild passerine birds. *Environmental Research*, **26**, 363, 1981.
5. Greger J.L.: Aluminum content of the american diet. *Food Technology*, **39**, 75, 1985.
6. Miller R.G., Kopfler F.C., Kelty K.C., et al.: The occurrence of aluminum in drinking water. *Am. Water Works Assoc.*, **77**, 84, 1984.
7. Kaehny W.D., Hegg A.P., Alfrey A.C.: Gastrointestinal absorption of aluminum from aluminum-containing antacids. *N. Engl. J. Med.*, **296**, 1389, 1977.
8. Alfrey A.C.: Aluminum metabolism in uremia. *Neurotoxicology*, **1**, 43, 1980.
9. Alfrey A.C., Hegg A., Craswell P.: Metabolism and toxicity of aluminum in renal failure. *Am. J. Clin. Nutr.*, **33**, 1509, 1980.
10. Mussi I., Calzaferri G., Buratti M., et al.: Behaviour of plasma and urinary aluminum levels in occupationally exposed subjects. *Int. Arch. Occup. Environ. Health*, **54**, 155, 1984.
11. Froment D.H., Buddington B., Miller N.L., Alfrey A.C.: Effect of solubility on the gastrointestinal absorption of aluminum from various aluminum compounds in the rat: *J. Lab. Clin. Med.*, **114**, 237, 1989.

12. Provan S.D., Yokel R.A.: Aluminum uptake by the in situ rat gut preparation. *J. Pharmacol. Exp. Ther.*, **245**, 928, 1988.
13. Froment D.H., Molitoris B.A., Buddington B., et al.: Site and mechanism of enhanced gastrointestinal absorption of aluminum by citrate. *Kidney Int.*, **36**, 978, 1989.
14. Molitoris B.A., Froment D.H., Mackenzie T.A., et al.: Citrate: A major factor in the toxicity of orally administered aluminum compounds. *Kidney Int.*, **36**, 949, 1989.
15. Ittel T.H., Buddington B., Miller N.L., Alfrey A.C.: Enhanced gastrointestinal absorption of aluminum in uremic rats. *Kidney Int.*, **32**, 821, 1987.
16. Gorsky J.E., Dietz A.A., Spencer H., et al.: Metabolic balance of aluminum studied in six men. *Clin. Chem.*, **25**, 1739, 1979.
17. Greger J.L., Baier M.J.: Excretion and retention of low or moderate levels of aluminum by human subjects. *Fd. Chem. Toxicol.*, **21**, 473, 1983.
18. Nolan C.R., DeGoes J.J., Mantini S.L., Alfrey A.C.: Gastrointestinal absorption of aluminum and lead in women ingesting calcium citrate. Submitted.
19. Adler A.J., Berlyne G.M.: Duodenal aluminum absorption in the rat: Effect of vitamin D. *Am. J. Physiol. Gastrointes. Liver Physiol.*, **12**, G209, 1985.
20. Mayor G.H., Sprague S.M., Hourani M.R., et al.: Parathyroid hormone-mediated aluminum deposition and egress in the rat. *Kidney Int.*, **17**, 40, 1980.
21. Cannata J.B., Suarez S.C., Buesta V., et al.: Gastrointestinal aluminum absorption: Is it modulated by the iron-absorptive mechanism? *Proc. EDTA-ERA*, **21**, 354, 1985.
22. Kovalchik M.T., Kaehny W.D., Kegg A., et al.: Aluminum kinetics during hemodialysis. *J. Lab. and Clin. Med.*, **92**, 712, 1978.
23. Klein G.L., Ott S.M., Alfrey A.C., et al.: Aluminum as a factor in the bone disease of long-term parenteral nutrition. *Trans. Assoc. Am. Physicians*, **45**, 115, 1982.
24. Slanina P., Falkeborn Y.: Aluminum concentrations in the brain and bone of rats fed citric acid, aluminum citrate or aluminum hydroxide. *Fd. Chem. Toxicol.*, **22**, 391, 1983.
25. Alfrey A.C., LeGendre G.R., Kaehny W.D.: The dialysis encephalopathy syndrome. Possible aluminum intoxication. *N. Engl. J. Med.*, **294**, 184, 1976.
26. Alfrey A.C.: Dialysis encephalopathy syndrome. *Annu. Rev. Med.*, **2**, 93, 1978.
27. Kaehny W.D., Alfrey A.C., Holman R.E., et al.: Aluminum transfer during hemodialysis. *Kidney Int.*, **12**, 361, 1977.
28. Milliner D.S., Shinaberger J.H., Shuman P., Coburn J.W.: Inadvertent aluminum administration during plasma exchange due to aluminum contamination of albumin replacement solutions. *N. Engl. J. Med.*, **312**, 165, 1985.
29. Sedman A.B., Klein G.L., Merritt R.J., et al.: Evidence of aluminum loading in infants receiving intravenous therapy. *N. Engl. J. Med.*, **312**, 1337, 1983.
30. Quagliaro D.A., Geraci V.A., Dwam R.E., et al.: Aluminum in albumin for injection. *J. Parentheral Sci. Tech.*, **42**, 187, 1988.
31. Victor R., Chan A.K., Mattoon M.: Aluminum contamination in albumin solutions from glass storage. *Transfusion*, **28**, 290, 1988.
32. Recker R.R., Blotcky A.J., Leffler J.A., et al.: Aluminum Evidence for aluminum absorption from the gastrointestinal tract and bone deposition by aluminum carbonate ingestion with normal renal function. *J. Lab. Clin. Med.*, **90**, 810, 1977.
33. Alfrey A.C.: Aluminum. *Adv. Clin. Chem.*, **23**, 69, 1983.
34. Bakir A.A., Hryhorczuk D.O., Berman E., Dunea G.: Acute fatal hyperaluminemic encephalopathy in undialyzed and recently dialyzed uremic patients. *Trans. Am. Soc. Artif. Intern. Organs*, **32**, 171, 1986.
35. Milliner D.S., Nebeker H.G., Ott S.M., et al.: Use of the deferoxamine infusion test in the diagnosis of aluminum-related osteodystrophy. *Ann. Intern. Med.*, **101**, 775, 1984.
36. Hodsman A.B., Sherrard D.J., Alfrey A.C., et al.: Bone aluminum and histomorphometric features of renal osteodystrophy. *Clin. Endocrinol. Metab.*, **54**, 539, 1982.
37. Hodsman A.B., Sherrard D., Wong E.G.C., et al.: Vitamin D-resistant osteomalacia in hemodialysis patients lacking secondary hyperparathyroidism. *Ann. Intern. Med.*, **94**, 629, 1981.
38. Ott S.M., Maloney N.A., Coburn J.W., et al.: The prevalence of bone aluminum deposition in renal osteodystrophy and its relation to the response to calcitriol therapy. *N. Engl. J. Med.*, **307**, 709, 1982.
39. O'Hare J.A., Murnaghan D.J.: Reversal of aluminum-induced hemodialysis anemia by low-aluminum dialysate. *N. Engl. J. Med.*, **306**, 654, 1982.
40. Siem (1866): Quoted in Doeilken: Ober die Wirkung des Aluminum, unter besonder Berucksichtigung der durch das Aluminum versurachtenlasionen im Zentralnervensystem. *Nauym-Schmiedelbergs Arch. Exp. Pathol. Pharmak.*, **40**, 58, 1897.
41. Scherp H.W., Church C.F.: Neurotoxic action of aluminum salts. *Proc. Soc. Exp. Biol. Med.*, **36**, 851, 1937.
42. Crapper D.R., Kirshnan S.S., Quittkat S.: Aluminum, neurofibrillary degeneration and Alzheimer's disease. *Brain*, **99**, 67, 1976.
43. Klatzo I., Wisniewski H., Stretcher E.: Experimental production of neurofibrillary degen-

eration. 1. Light microscopic observations. *J. Neuropathol.*, **24**, 187, 1965.
44. Petit T.L., Biederman G.B., McMullen P.A.: Neurofibrillary degeneration, dendritic dying back, and learning-memory deficits after aluminum administration: implications for brain aging. *Exp. Neurol.*, **67**, 152, 1980.
45. Terry R.D., Pena C.: Experimental production of neurofibrillary degeneration. 2. Electron microscopy, phosphatase histochemistry and electron probe analysis. *J. Neuropathol.*, **24**, 200, 1965.
46. Wisniewski H.M., Sturman J.A., Shek W.: Aluminum chloride-induced neurofibrillary changes in the developing rabbit: A chronic animal model. *Ann. Neurol.*, **8**, 479, 1980.
47. De Boni U., Otvos A., Scott J.W., et al.: Neurofibrillary degeneration induced by systemic aluminum. *Acta Neuropathol.* (Berlin), **35**, 285, 1976.
48. Wisniewski H.M., Iqbal K.: Aluminum-induced neurofibrillary changes: Its relationship to senile dementia of the Alzheimer's type. *Neurotoxicology*, **1**, 121, 1980.
49. McLaughlin A.I.G., Kazantzis G., King E., et al.: Pulmonary fibrosis and encephalopathy associated with the inhalation of aluminum dust. *Br. J. Ind. Med.*, **19**, 253, 1962.
50. Lipresle J., Duckett S., Galle P., et al.: A case of aluminum encephalopathy in man. *Cr Seances Soc. Bio. Ses. Fil.*, **169**, 282, 1975.
51. Alfrey A.C., Mishell M.M., Burks J., et al.: Syndrome of dyspraxia and multifocal seizures associated with chronic hemodialysis. *Trans. Am. Soc. Artif. Intern Organs*, **18**, 257, 1972.
52. Berkseth R., Mahowald M., Anderson D., et al.: Dialysis encephalopathy (DE): Diagnostic criteria and epidemiology of 39 patients. *Am. Soc. Nephrol.*, **11**, 36A, 1978.
53. Dunea G., Mahuraker S.D., Mamdani B., et al.: Role of aluminum in dialysis dementia. *Ann. Intern. Med.*, **88**, 502, 1978.
54. Flendrig J.A., Kruis H., Das H.A.: Aluminum intoxication: The cause of dialysis dementia? *Proc. Eur. Dial. Transplant. Assoc.*, **13**, 355, 1976.
55. Galle P., Chatel M., Berry J.P., et al.: Encephalopathie myoclonique progressive des dialyses: Presence d'aluminum et forte concentration dans les lisosomes des cellules cerebrates. *Nouv. Presse Med.*, **8**, 4091, 1979.
56. McDermott J.R., Smith A.I., Ward M.K., et al.: Brain-aluminum concentration in dialysis encephalopathy. *Lancet*, **1**, 901, 1978.
57. Schroeder M.T.: Dialysis encephalopaty. *Arch. Intern. Med.*, **139**, 510, 1979.
58. Rozas V.V., Port K.F., Rutt W.M.: Progressive dialysis encephalopathy from dialysate aluminum. *Arch. Intern. Med.*, **138**, 1375, 1978.
59. Parkinson I.S., Ward M.K., Feest T.G., et al.: Fracturing dialysis osteodystrophy and dialysis encephalopathy. An epidemiological survey. *Lancet*, **1**, 406, 1979.
60. Platts M.M., Goode G.C., Hislop J.S.: Composition of the domestic water supply and the incidence of fracture and encephalopathy in patients on home dialysis. *Br. Med. J.*, **2**, 657, 1977.
61. Cartier F., Allain P., Garv J., et al.: Encephalopathie myclonique progressive des dialyses. Role de l'eau utilisee pour l'hemodialyse. *Nouv. Presse Med.*, **7**, 97, 1978.
62. Ackrill P., Ralston A.J., Day J.P., et al.: Successful removal of aluminum from patient with dialysis encephalopathy. *Lancet*, **2**, 692, 1980.
63. Milne F.J., Sharf B., Bell P., et al.: The effect of low aluminum water and desferrioxamine on the outcome of dialysis encephalopathy. *Clin. Nephrol.*, **20**, 202, 1983.
64. Ackrill P., Ralston A.J., Day J.P.: The role of desferrioxamine in the treatment of dialysis encephalopathy. *Kidney Int.*, **29** (suppl. 18), 104, 1986.
65. Kirschbaum B.B., Schoolwerth A.C.: Acute aluminum toxicity associated with oral citrate and aluminum-containing antacids. *Am. J. Med. Sci.*, **297**, 9, 1989.
66. Sedman A.B., Miller N.L., Warady B.A., et al.: Aluminum loading in children with chronic renal failure. *Kidney Int.*, **26**, 201-204, 1984.
67. Burks J.S., Alfrey A.C., Huddlestone J., et al.: A fatal encephalopathy in chronic hemodialysis patients. *Lancet*, **1**, 764, 1976.
68. Pierides A.M., Edwards W.G. Jr., Cullum U.X Jr., et al.: Hemodialysis encephalopathy with osteomalacic fractures and muscle weakness. *Kidney Int.*, **18**, 115, 1980.
69. Chan Y., Alfrey A.C., Posen S., et al.: The effect of aluminum on normal and uremic rats: Tissue distribution, vitamin D metabolites and quantitative bone histology. *Calcif. Tissue Int.*, **35**, 344, 1983.
70. Goodman W.G., Henry D.A., Horst R., et al.: Parenteral aluminum administration in the dog. II. Induction of osteomalacia and effect on vitamin D metabolites. *Kidney Int.*, **25**, 370, 1984.
71. Sedman A.B., Alfrey A.C., Miller N.L., Goodman W.G.: Tissue and cellular basis for impaired bone formation in aluminum-related osteomalacia in the pig *J. Clin. Invest.*, **79**, 86, 1987.
72. Llach F., Felsenfeld A.J., Coleman M.D., et al.: The natural course of dialysis osteomalacia. *Kidney Int.*, **29** (suppl. 18), S74, 1986.
73. Chan Y-L., Furlong T.J., Cornish C.J., et al.: Dialysis osteodystrophy: A study involving 94 patients. *Medicine*, **64**, 296, 1985.
74. Dahl E., Nordal K.P., Attramadal A., et al.: Renal osteodystrophy in predialysis patients without stainable bone aluminum. *Acta Med. Scand.*, **224**, 157, 1988.
75. Maloney N.A., Ott S., Alfrey A.C., et al.: Histologic quantitation of aluminum in iliac bone

from patients with renal failure. *J. Lab. Clin. Med.*, **99**, 206, 1982.

76. Short A.L.K., Winney R.J., Robsop J.S.: Reversible microcytic hypochromic anaemia in dialysis patients due to aluminum intoxication. *Proc. Eur. Dial. Transplant Assoc.*, **17**, 226, 1980.
77. Touam M., Martinez F., Lacour B., et al.: Aluminum-induced, reversible microcytic anemia in chronic renal failure: Clinical and experimental studies. *Clin. Nephrol.*, **19**, 295, 1983.
78. Kaiser L., Burnatowska-Hledin M.A., Schwartz K.A., et al.: Microcytic anemia secondary to intraperitoneal aluminum in the rat (abst). *Kidney Int.*, **25**, 194, 1984.
79. Seibert F.B., Wells H.G.: The effect of aluminum on mammalian blood and tissue. *Arch. Pathol.*, **8**, 230, 1929.
80. Casati S., Castelnovo C., Campise M., Ponticelli C.: Aluminum interference in the treatment of haemodialysis patients with recombinant erythropoietin. *Nephrol. Dial. Transplant*, **5**, 441, 1990.
81. Seyle H.: *Calciphylaxis*. University of Chicago Press, Chicago, 1962.
82. Johannsson O., Perrault L., Savoie L., Tuchweber B.: Action of various metallic chlorides on calcaemic and phosphataemia. *Br. J. Pharmac. Chemoter.*, **33**, 91, 1986.
83. Martin R.B.: The chemistry of aluminum as related to biology and medicine., *Clin. Chem.*, **32**, 1797-1806, 1986.
84. Mladenovic J.: Aluminum inhibits erythropoiesis in vitro. *J. Clin. Invest.*, **81**, 1661, 1988.
85. Zhu J.M., Huffer W., Alfrey A.C.: Effect of aluminum on bone matrix inductive properties. *Kidney Int.*, **38**, 1141, 1990.
86. Fournier A., Moriniere P., Sebert J., et al.: Calcium carbonate, an aluminum-free agent for control of hyperphosphatemia, hypocalcemia and hyperparathyroidism in uremia. *Kidney Int.*, **29** (suppl. 18), S114, 1986.
87. Sheikh M.S., Maquire J.A., Emmett M., et al.: Reduction of dietary phosphorus absorption by phosphorus binders: A theoretical, in vitro and in vivo study. *J. Clin. Invest.*, **83**, 66, 1989.
88. Hudson G.A., Milne F.J., Meyers A.M., et al.: Treatment of dialysis fracturing bone disease. *Kidney Int.*, **17**, 532, 1980.
89. Vargas J.H., Klein G.L., Ament M.E., et al.: Metabolic bone disease of total parenteral nutrition: Course after changing from casein to amino acids in parenteral solutions with reduced aluminum content. *Am. J. Clin. Nutr.*, **48**, 1070, 1988.
90. Nebeker H.G., Milliner D.S., Ott S.M., et al.: Aluminum-related osteomalacia: Clinical response to desferrioxamine. *Kidney Int.*, **25**, 173, 1984.
91. Ciancioni C., Poignet J.L., Mauras Y., et al.: Plasma aluminum and iron kinetics in hemodialyzed patients after IV infusion of desferrioxamine. *Trans. Am. Soc. Artif. Intern. Organs*, **30**, 479, 1984.
92. Ciancioni C., Poignet J.L., Naret C., et al.: Concomitant removal of aluminum and iron by haemodialysis and haemofiltration after desferrioxamine intravenous infusion. *Proc. EDTA-ERA*, **21**, 469, 1984.
93. Molitoris B.A., Alfrey P.S., Miller N.L., et al.: Efficacy of intramuscolar and intraperitoneal deferoxamine for aluminum chelation. *Kidney Int.*, **31**, 986, 1987.
94. Kurokawa K., Ogura Y., Ono T., et al.: Deferoxamine in aluminum induced bone disease in dialysis patients. VIth UOEH International Sym. and the IIIrd COMTOX on Bio- and Toxic Kinetics of Metals. Kitakyushu City, Japan, July 27, 1986.
95. Veis J.H., Contiguglia R., Klein M., et al.: Mucormycosis in deferoxamine treated patients on dialysis. *Ann. Inter. Med.*, **107**, 258, 1987.
96. Brock J.H.: The effect of desferrioxamine on the growth of staphylococus aureus, yersinia enterocolitica and streptococus faecalis in human sera: uptake of deferrioxamine-bound iron. *FEMS Microbiol. Letters*, **20**, 439, 1983.

Alzheimer's Disease and the Aluminium Hypothesis

J.A. Edwardson, I.N. Ferrier, F.K. McArthur, I.G. McKeith,
I. McLaughlin, C.M. Morris, S.A. Mountfort, A.E. Oakley,
G.A. Taylor, M.K. Ward, J.M. Candy

1. Introduction

While genetic linkage analyses have shown that familial Alzheimer's disease is, in some cases, due to autosomal dominant inheritance of a gene defect on chromosone 21, a genetic cause has not been established for the much larger proportion of apparently sporadic cases which occur in the elderly population. A recent review [1] has listed over twenty risk factors which have been implicated in the aetiopathogenesis of Alzheimer's disease, although the evidence for most of these is slight. Since the original observations that neurofibrillary changes occur in animals injected with aluminium salts,[2,3] this element has been repeatedly implicated as a risk factor for Alzheimer's disease on the basis of neuropathological, experimental and epidemiological studies. However, despite these numerous investigations, the evidence is not conclusive and, given the public health implications associated with the aluminium hypothesis, the subject remains highly controversial.

It is relevant to consider the likely criteria which would have to be fulfilled by a candidate environmental toxin which could account for this disorder. Thus: (i) Given the worldwide distribution of Alzheimer's disease it is reasonable to suppose that the putative toxin should have an equally wide geographical distribution. (ii) The high prevalence rates of Alzheimer's disease in elderly populations suggest that such a factor is likely to be ubiquitous in the environment. Reported geographical variations in the incidence rates for Alzheimer's disease should be consistent either with differences in exposure or the response to exposure in the populations concerned. (iii) Selective uptake by relevant neuronal populations would provide a convincing explanation for the regional vulnerability of the brain areas affected in Alzheimer's disease. (iv) The age-related incidence of Alzheimer's disease and the considerable time span which appears to be involved in the formation of senile plaques and neurofibrillary tangles allow the possibility that the putative environmental agent accumulates equally slowly in the nervous system – although the «threshold model» for neurodegenerative disorders, first developed for Parkinson's disease, also allows for an early, discrete insult to produce symptoms after a considerable latent period during which the compensatory capacity of surviving neurones is reduced by age-related attrition. (v) The putative factor should have neurotoxic actions capable of producing the appropriate cellular and biochemical lesions, eg neurofibrillary changes and β-amyloid deposition. (vi) The consistent and specific association of such a risk factor with plaques and tangles, the hallmark neuropathological lesions of Alzheimer's disease would also be significant. (vii) Since only a minority of aged individuals develop Alzheimer's disease, it would be relevant to show that the putative factor interacts with genetic or other constitutional factors in a way consistent with this susceptibility. The extent to which aluminium fulfills these criteria as a potential neurotoxic agent in the aetiopathogenesis of Alzheimer's disease will now be considered.

2. Aluminium in the Environment

After oxygen and silicon, aluminium is the most abundant element, consistuting approximately 8% of the earth's outer crust. The chemistry of aluminium in the environment is complex and poorly understood [4] but the major, naturally occurring forms present in rocks, clays and soils are highly insoluble and bioavailable forms are largely due to human activity – the manufacture of metallic aluminium or salts, and the consequences of acid rain. The total body burden of aluminium in adult man has been estimated to be of the order 30-50 mg; approximately 25% being present in the lungs, probably as a result of atmospheric intake, 25% in soft tissues and the remainder in bone.[5] This low body content, which is only twice the average daily intake of aluminium,[6] reflects the insolubility of most naturally occurring forms of aluminium, the considerable gastro-intestinal barrier to the absorption of bioavailable species, and the efficiency of renal excretion of absorbed aluminium in healthy individuals.

3. Uptake and Distribution of Aluminium in Brain

In blood, aluminium appears to be mainly, if not exclusively, bound to the iron transporting protein transferrin. Although there is no suitable isotope of aluminium for use in studies on blood-brain permetability in man, we have used ^{67}Ga as a marker for brain uptake of aluminium in the rat.[7] It was found that ^{67}Ga bound to plasma transferrin and that it entered the brain unidirectionally, with a permeability of 2.48×10^{-6} ml/min/g. The autoradiographic distribution of ^{67}Ga administered *in vivo* showed a discrete regional distribution with high levels in cerebral cortex, hippocampus amygdala and, septum, areas which also had high densities of transferrin receptors, as evidenced by the binding of ^{125}I-Fe-transferrin *in vitro*. Assuming a similar low blood-brain permeability for aluminium we have calculated[7] that the human brain will accumulate approximately 8 μg per year, an amount consistent with the observed content of aluminium after lifetime exposure.[8] Confirmation that transferrin-mediated brain uptake of aluminium may explain the slow rate of accumulation and selective distribution of this element was obtained in a *post-mortem* study of the brains of patients with chronic renal failure.[9] These patients are unable to excrete aluminium and, in addition, are often treated with large doses of aluminium containing phosphate binders to prevent hyperphosphataemia. Imaging secondary ion-mass spectrometry (SIMS) has demonstrated the presence of numerous focal accumulations of aluminium, which correspond largely with the distribution of pyramidal neurones. These cell-like profiles also correlated with the distribution of transferrin receptors in the forebrain (Fig. 1). The brain regions which show selective accumulation of ^{67}Ga in the rat and aluminium in patients with chronic renal failure coincide with areas that are selectively vulnerable in both dialysis encephalopathy and, to a large extent, in Alzheimer's disease. We suggest therefore that this selective vulnerability reflects, at least in part, the distribution of transferrin receptors which occur in high densities on the most metabolically active neurones – cells which have the highest requirements for iron in the synthesis of respiratory chain enzymes.[10]

Studies on the brain content of aluminium in Alzheimer's disease have not shown a consistent elevation in levels and this has been used to argue against a pathogenic involvement for aluminium. However, our investigations on patients with chronic renal failure using imaging SIMS have revealed extensive intraneuronal accumulation in cases which fall within the normal range of aluminium content as judged by atomic absorption spectroscopy (Fig. 2). A preliminary quantitative investigation using SIMS has indicated that these intracellular concentrations can exceed 500 ppm (Candy et al., unpublished results) and it is thus relevant to consider whether pathological changes are associated with intraneuronal accumulation.

4. β-Amyloid Deposition in Chronic Renal Failure

Studies in patients with dialysis encephalopathy have shown the presence of spongiform changes in the neuropil[11, 12] and one investigation reported the presence of both senile plaques and neurofibrillary tangles in both encephalopathic and non-encephalopathic dialysis patients.[13] In a study of patients with

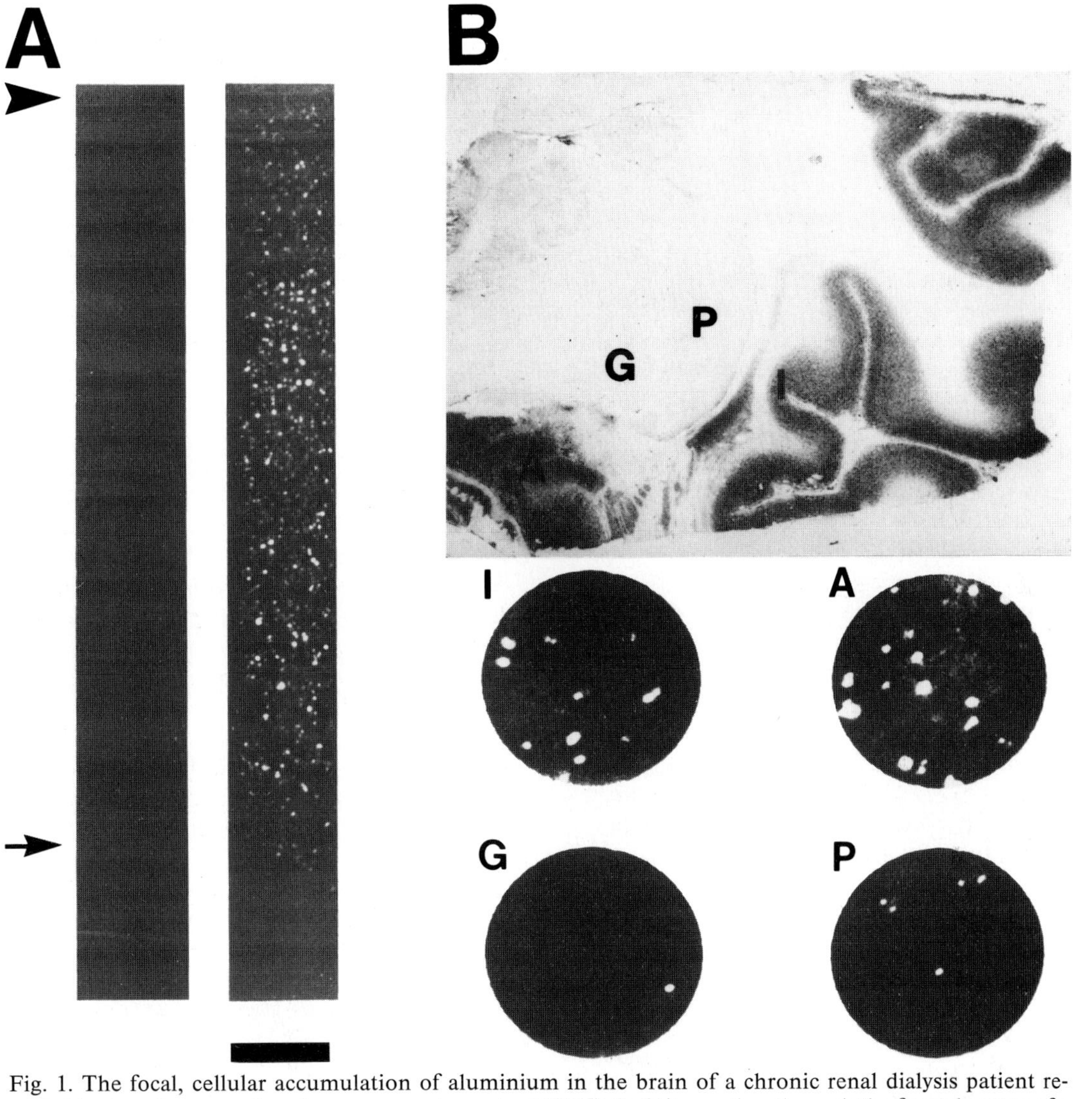

Fig. 1. The focal, cellular accumulation of aluminium in the brain of a chronic renal dialysis patient revealed by imaging secondary ion mass spectrometry (SIMS). In (A) a section through the frontal cortex of a dialysis patient (right) is compared with a matched control case (left). The upper and lower arrows mark the pial surface and grey/white matter boundary. Scale bar represents 400 µm. In (B) the distribution of transferrin receptor sites, revealed by autoradiography with [^{125}I]-Fe-transferrin, in the basal forebrain of a dialysis patient is shown (upper). Areas with high densities of transferrin receptors such as insula cortex (I) or amygdala (A) show high densities of aluminium containing cells using SIMS, compared with areas such as putamen (P) or globus pallidus (G) which are low in transferrin receptors.

chronic renal failure we have recently found [14] nuclear immunostaining with an antibody to the N-terminal region of the β-amyloid precursor protein in 14 out of 15 cases studied (Fig. 3 A). Five of these patients had amorphous (non-neuritic) senile plaques in the frontal, temporal, parietal and occipital cortices, on staining with Kings silver carbonate or antisera

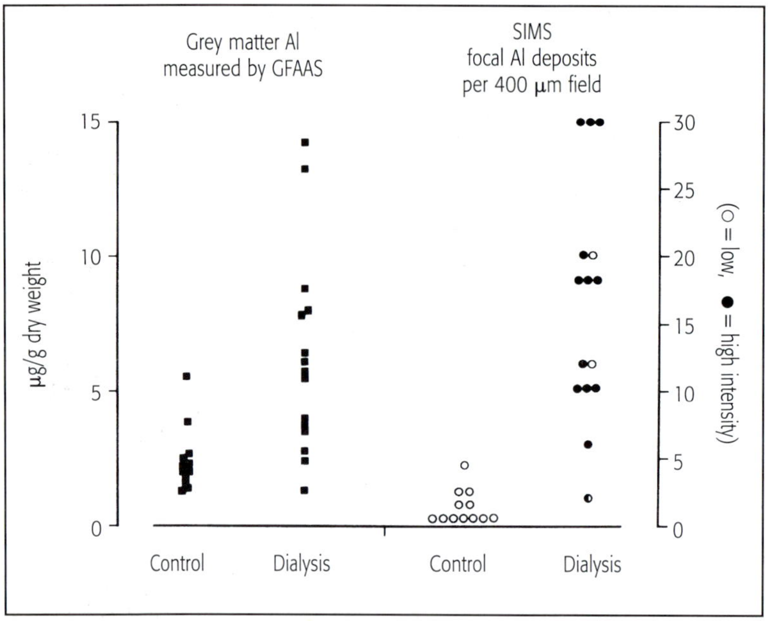

Fig. 2. Comparison between the concentration of aluminium in the cerebral cortex determined using graphite furnace atomic absorption spectrometry and the number and intensity of the focal aluminium accumulations present in the cerebral cortex determined using secondary ion mass spectrometry in chronic renal dialysis patients.

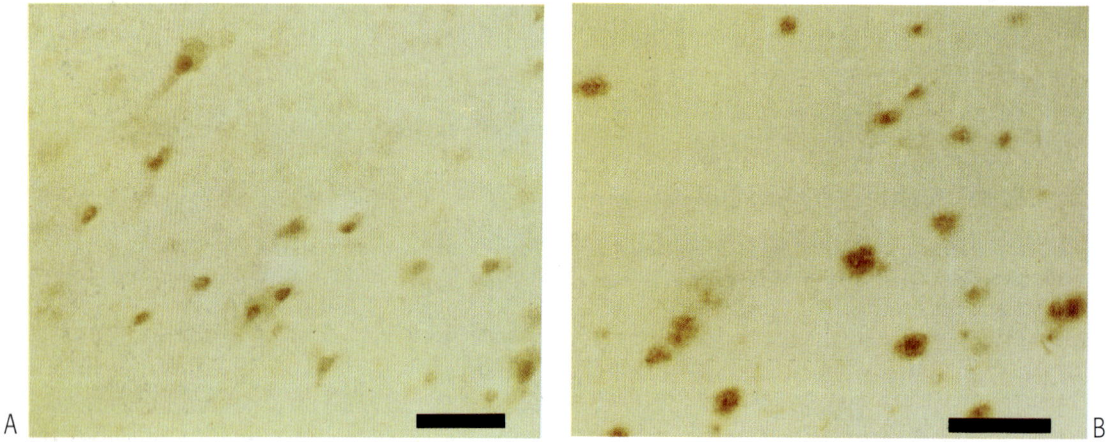

Fig. 3. A) Shows β/A4 amyloid precursor protein immunostaining (GP3 antibody) associated with pyramidal cell nuclei and cytoplasm in the cortex of a chronic renal dialysis patient. No staining is evident in age matched control cases. B) Shows the presence of A4 immunostained «immature» senile plaques in the cerebral cortex of the same chronic renal dialysis patient. The scale bar in A represents 40 μm and in B 500 μm.

against the β-peptide. The plaques were identical (Fig. 3B) to the «immature» plaques seen in Alzheimer's disease, Down's syndrome and some elderly mentally normal individuals,[15] and at the ultrastructural level consisted of a loose meshwork of fibrillary aggregates with an individual fibre diameter of 30-40 Å. Very few core containing plaques were observed but other studies have estimated that it takes up to three decades to form mature core-containing plaques[16] and the mean period of dialysis in the patients with renal failure was 4-5 years. Although dialysis patients suffer from many metabolic and other disorders and the β-amyloid precursor is a cell-stress related protein, it is striking that the intracellular staining for this protein occurred predominantly in the pyramidal neurones, which selectively accumulate aluminium. Neurofibrillary tangles were not observed in the dialysis cases but, again, these may require a considerable timespan to develop since they occur in Down's syndrome long after the first evidence of β-amyloid deposition. Although many different effects of aluminium at the cellular level have been described[17] this study on dialysis cases suggest that aluminium may induce some Alzheimer type neuropathological changes through mechanisms which involve transferrin-mediated uptake by neurones increased expression or decreased turnover of the β-amyloid precursor protein and extracellular deposition of fibrillary aggregates of β-amyloid, perhaps as a consequence of altered processing of the β-amyloid precursor protein.[18]

5. Association of Aluminium with Plaques and Tangles

During the course of attempts to isolate and characterize the amyloid protein in senile plaque cores, we observed that significant amounts of non-proteinaceous material were associated with highly purified preparations. Electron microprobe X-ray microanalysis revealed the presence of Al and Si in the central region of plaque cores and the presence of these elements as amorphous aluminosilicates was indicated by solid state ^{27}Al nuclear magnetic resonance spectroscopy.[19,20,21] Focal deposits of Al and Si were observed as consistent feature of plaque cores in presenile and senile Alzheimer's disease, Down's syndrome and normal elderly individuals (Fig. 4). The presence of such deposits *in situ* has been confirmed using both energy and wave-length dispersive electron probe microanalysis, scanning proton microprobe X-ray analysis and SIMS. Similar deposits have not been found associated with other neuropathological inclusion bodies such as Pick bodies or corpora amylaceae or with other forms of amyloid deposit including serum A amyloidosis of kidney tissue, immunoglobulin light chain amyloidosis of spleen and congophilic angiopathy of cerebral blood vessels which involves deposition of the β-amyloid protein.[22] Although several groups have failed to replicate these findings, there are considerable technical problems associated with such procedures, including: (i) difficulties in identifying the plaque core where a secondary electron detector or other methods for topographical mapping are not

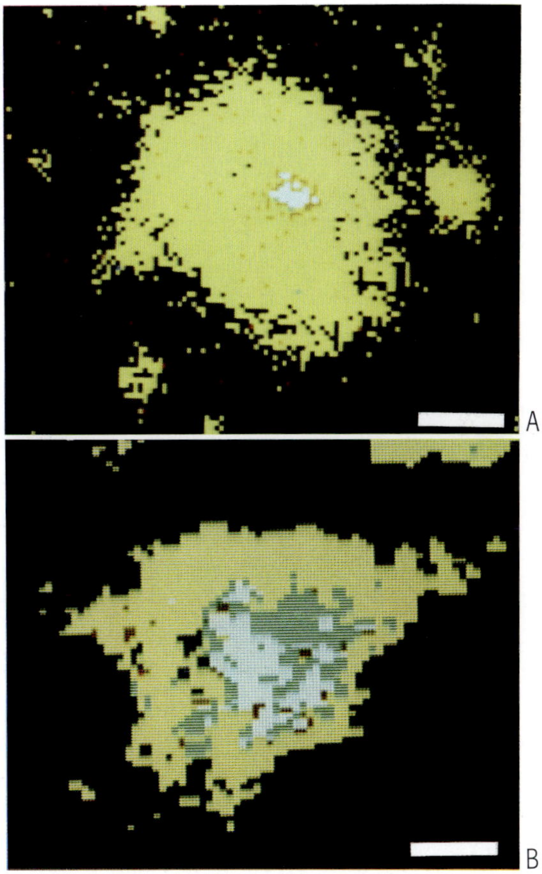

Fig. 4. The presence of aluminium and silicon in the core of senile plaques in (A) Alzheimer's disease and (B) Down's syndrome, demonstrated using electron-induced energy dispersive X-ray microanalysis. Sections of cerebral cortex were stained with silver (yellow) to identify plaques and the white areas indicate the co-localization of silicon (blue) and aluminium (red) in the central core region. The scale bar in A represents 10 μm and in B 4 μm.

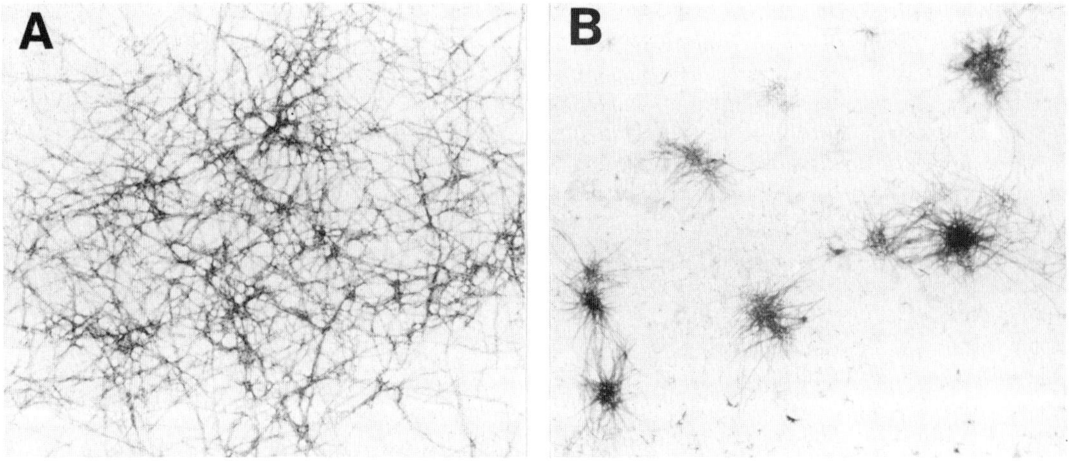

Fig. 5. Fibrillary aggregates of substance P in A) distilled water, B) an amorphous aluminosilicate suspension in distilled water. Note in A) the individual random appearance of the fibrils and in B) fibrillary aggregates with a stellate appearance. Energy dispersive X-ray microanalysis using scanning transmission electron microscopy revealed the presence of aluminosilicate in the centre of each stellate aggregate. The calibration bars represent 700 nm in A and 1400 nm in B.

used; (ii) failure to recognize that if Al and Si are focally co-localized in the most *central* region of the plaque core at least half of sections through plaques will be above or below this level and (iii) use of electron beam energies insufficient (< 15 KeV) to penetrate the core matrix and reach the site of Al deposition.

It is of interest that the amyloid fibrils in senile plaque cores are present in a radiating «star-burst» array, in contrast to the amorphous or laminar arrangements of fibrillary aggregates in other forms of amyloidosis. Previously, we have shown that several peptides form linear, polymeric aggregates in water or sodium chloride.[23] We have recently found that, in the presence of aluminosilicates, peptides which normally form amorphous tangles of fibrillary aggregates (Fig. 5 A) are seen arranged in dense, radiating clusters which resemble the «star-burst» array of mature plaque cores (Fig. 5 B). These observations suggest that aluminium or aluminosilicates may account for a structural arrangement of β-amyloid in plaque cores which could render it less susceptible to removal by microglial cells or other phagocytic elements.

It has also been reported that the aluminium content of neurofibrillary tangle bearing neurones is increased in Alzheimer's disease[24] and more recent observations using laser microprobe mass analysis suggest that intracellular concentrations associated with tangles are much lower than we have observed in brains of dialysis patients.[25] While the presence of aluminium in both plaques and tangles does not necessarily indicate an aetiopathogenic role, these findings must be considered in the light of other evidence such as the precocious deposition of β-amyloid in renal dialysis cases, discussed previously, and epidemiological evidence that the relative risk for Alzheimer's disease is greater in populations exposed to higher levels of aluminium in water supplies.

6. Epidemiological Studies

There are now five different epidemiological studies which implicate aluminium in water supplies as a significant risk factor in the aetiopathogenesis of Alzheimer's disease. Aluminium sulphate is used extensively as a flocculation agent to remove organic substances such as humic acid which cause the brown, peaty discolouration of surface waters. Soluble forms of aluminium are also released from rocks and soils under acidic conditions and acid rain is a significant factor in the mobilization of aluminium into rivers and lakes. It seems likely that fluoride will form insoluble complexes with reactive forms of aluminium and could thus reduce the absorption of aluminium from the gastrointestinal tract. Still and Kelly compared hospital admission rates for primary dementia in two counties which had high and low fluoride content in their water supplies and found that admissions in the high fluoride district were only 19% of the rate in the low fluoride area, a difference which they attributed to the effects of fluoride on aluminium absorption.[26] Two Norwegian studies have shown a significant positive correlation between concentrations of aluminium in drinking water and mortality rates for dementia, determined from death certification, and also long-stay bed occupancy by dementia sufferers.[27,28] In the most extensive study of this issue to-date, Martyn et al.[29] determined incidence rates for probable pre-senile Alzheimer's disease using clinical information provided in letters of referral to CT brain-scanning units in 88 county districts within 7 regions in England and Wales. A significant, increased relative risk of 1.7 was shown in areas where the aluminium content of water exceeded 110 μg/litre and some evidence of a dose response was observed (Table I) with patients under 65 years of age where the likelihood of referral by clinicians for a CT scan is greatest. A further, recent study in SW France, reported in preliminary form[30] and which involved clinical screening of elderly populations has indicated a significantly elevated relative risk of 4.9 for Alzheimer's disease in areas where the aluminium concentrations exceeded 110 μg/litre. While these diverse studies can be seriously criticized on a number of grounds including: the inadequacy of death certification to provide accurate mortality rates for dementia; the accuracy of diagnoses based upon clinical information in referral letters; possible regional differences in CT scanning rates; possible movement of patients between different areas etc — all of these confounding factors would tend to lessen the chances of a positive association. However, aluminium in drinking water accounts, at the most, for only 10% of the average daily aluminium intake where there are heavily-treated water supplies, and much less in other areas, and Birchall has suggested a possible, alternative explanation for these epidemiological findings.[31] He has pointed out that there is a broad inverse correlation between the concentration of silicon, in the form of silicic acid, and that of aluminium in water supplies. Since high levels of silicic acid may promote the formation of aluminosilicate species within the gastrointestinal tract and prevent the absorption of aluminium from all dietary sources, it is possible that the geographical correlation observed in the epidemi-

Table I. **Relative risk (95% CI) of probable Alzheimer's disease and dementia from other causes and epilepsy in patients aged < 65 years in relation to aluminium content of water supplies in district of residence** (data from Martyn et al., ref. 29).

Al content (μg/l)	Probable Alzheimer's disease (n = 307)	Other causes of dementia (n = 372)	Late-onset epilepsy (n = 2461)
0 - 40	1	1	1
20 - 40	1.4 (1.0-2.2)	1.2 (0.8-1.7)	1.0 (0.8-1.1)
50 - 70	1.4 (1.0-2.2)	1.1 (0.8-1.6)	0.9 (0.8-1.1)
80 - 100	1.6 (1.0-2.5)	1.2 (0.8-1.8)	1.0 (0.9-1.2)
> 110	1.7 (1.1-2.7)*	1.2 (0.8-1.8)	1.0 (0.8-1.1)

* Significantly increased relative risk.

ological studies may reflect levels of silicic acid and decreased aluminium uptake, rather than levels of aluminium residuals *per se*. This important hypothesis concerning the possible protective effects of silicic acid is currently being tested.

Other dietary sources of aluminium in forms which are possibly bioavailable include food additives and a few naturally occurring substances, such as tea, which are rich in aluminium. Levels of aluminium in brewed tea are commonly in the range 2-6 mg/litre[32] which is at least an order of magnitude greater than the highest levels in water in the epidemiological studies reviewed above. However, study of the speciation of aluminium in brewed tea using molecular size fractionation and ion-exchange techniques has shown that 96% of the filtrable fraction has a molecular size greater than 10,000, suggesting that it may be present in high molecular weight complexes, and even after acidification to pH 2.0 and incubation at 37 °C for two hours (to mimic stomach conditions) 51% was retained by an ultrafiltration membrane with 10,000 molecular weight cut-off.[33] Thus, it is likely that polyphenolic or other organic compounds in tea may complex with aluminium and reduce its bioavailability.

Dietary intake of aluminium from foodstuffs is small compared with the amounts consumed through the use of aluminium-containing antacids which may provide 50-1000 mg/day of elemental aluminium. The use of antacids as a risk factor for Alzheimer's disease has been examined in several case-control studies and the results have proved negative.[34-37] However, two of these studies are small and do not have sufficient statistical power to detect such a relationship, and one of them provides evidence of a significant recall bias by informants – a likely problem in all studies of this type. Furthermore, since the absorption of aluminium occurs at least partly under the acidic conditions of the stomach and duodenum, the speciation of aluminium hydroxide to Al^{3+} will be prevented to a considerable extent by antacid use and the increases in aluminium absorption may not be directly proportional to the size of the dose.

Industrial exposure to aluminium dusts and fumes have also been investigated. There is evidence of pulmonary fibrosis and encephalopathy following massive occupational exposure to aluminium dust[38] and a recent study has provided evidence of cognitive impairment in miners exposed prophylactically over long periods (6m-36yrs) to a mixture of finely powdered aluminium and aluminium oxide to prevent silicotic lung disease.[39] In this study there were no significant differences between exposed and non-exposed groups in reported diagnoses of neurological disorder but dose-related impairments in cognitive function were observed using three tests of cognitive state. Given the large numbers of subjects exposed, and the considerable duration of exposure, it is noteworthy that there were no significant differences in the prevalence of dementia or other neurological disorder between the exposed and non-exposed groups. The direct absorption of inhaled aluminium from the lungs has been discussed but does not appear likely except for highly soluble aluminium compounds.[40] It is possible that inhaled aluminium could be transported to the gastro-intestinal tract and then absorbed but even this does not appear likely except with massive chronic exposures. We have recently measured whole blood levels of aluminium in 'scene-of-crime' police officers who have regularly used finely milled aluminium powder for finger printing during periods of employment up to 17 years.[41] Although serum aluminium is rapidly excreted by the kidneys in normal healthy individuals, it seems probable that the aluminium associated with red blood cells is incorporated during erythropoiesis and that whole blood assays should reflect exposure over a longer period. The mean whole blood aluminium concentration in 27 officers after a 3 day duty period (0.42 μmol/l) was not different to that measured after a similar period off-duty (0.43 μmol/l) or values obtained from control subjects, indicating that there is little or no absorption of aluminium following this type of exposure. This may explain why occupational exposure to aluminium does not appear to have been identified as a risk-factor in any of the epidemiological case-control studies on risk factors for Alzheimer's disease which have been reported to date.

7. Interaction with Genetic or Other Factors

If aluminium does play a contributory role in the aetiopathogenesis of Alzheimer's disease, the basis of the individual susceptibility to this disorder is, clearly, of major importance. The presence of aluminosilicates in plaque cores from patients with Down's syndrome does not necessarily argue against the «aluminium hypothesis» since there is other evidence for abnormal mineral homeostasis in trisomy 21, such as the increased incidence of calcification of the basal ganglia.[42, 43] Despite the fact that almost all renal dialysis patients show evidence of intraneuronal accumulation of aluminium and increased intraneuronal staining for the β-amyloid precursor, only a third of such cases had extracellular amyloid deposits in the form of immature senile plaques. This suggests that other factors, including genetic influences, may determine susceptibility to aluminium-induced neuropathological changes, and this view is supported by the evidence for a link between the HLA-B40 antigen and susceptibility to dialysis encephalopathy.[44] Individual susceptibility may not only be determined by the responsiveness of the central nervous system to aluminium. In a pilot study which we have carried out on the absorption of aluminium following an oral challenge in the presence of citrate (Table II) the rise in blood Al concentrations after 1 hour was significantly greater in younger (< 77 years) Alzheimer patients compared with a strictly aged-matched control group ($t = 0.0330$, $p < 0.05$, paired students t test, 1 tailed). Although a significant difference was not seen with an older group (> 77 years) of patients, this was because the response to challenge in the older control group was increased, indicating the possibility that in Alzheimer's disease there may be an early onset of this age-related increase in aluminium absorption. This increase in aluminium absorption following an oral challenge in the younger Alzheimer patients contrasts with the reduced gastrointestinal absorption of radioactive calcium which we have recently reported[45] and which occurs in the presence of normal plasma concentrations of parathyroid hormone and vitamin D metabolites. The reduced absorption of calcium appears to be a secondary, non-specific derangement since it also occurred in patients with multi-infarct dementia. Further evidence of a possible peripheral abnormality relating to aluminium metabolism in Alzheimer patients has been reported by Farrar et al..[46] This group used radioactive ^{67}gallium as a marker for aluminium and reported that gallium-transferrin binding was significantly lower in plasma from patients with Alzheimer's disease or Down's syndrome compared with normal age-matched controls. They suggest that a deficiency in aluminium binding to transferrin in Alzheimer's disease would leave more unbound aluminium which could be transported into brain. The

Table II. **Blood aluminium concentrations (μg/l ± SEM) before and 60 min after a drink containing 300 mg aluminium as citrate 67 mmol.**

Age	Group	n	Baseline	60 min	Increase
< 77	Control	10	9.8 ± 1.6	39.9 ± 4.8	30.1 ± 4.5
(66-76, mean 72)	Alzheimer	10	8.7 ± 0.9	103.1 ± 30.6	94.4 ± 30.3*
> 77	Control	10	6.8 ± 0.9	101.3 ± 70.1	94.5 ± 19.9
(77-89, mean 83)	Alzheimer	10	5.8 ± 0.4	79.8 ± 33.8	74.0 ± 33.6
All	Control	20	8.3 ± 1.0	70.6 ± 12.3	62.3 ± 12.4
(66-89, mean 77.5)	Alzheimer	20	7.2 ± 0.6	91.5 ± 22.4	84.3 ± 22.2

* Increase significantly different to control ($p < 0.05$, paired Students t-test, 1 tailed).

The chromatographic separation of transferrin-^{67}Ga in this study was not carried out under conditions in which the concentration of bicarbonate (which is essential for metal ion binding to transferrin acceptor sites) was maintained and this potentially important observation needs to be repeated under optimal conditions. The recent development of ^{26}Al accelerator mass spectrometry for biological investigations with a detection limit of approximately 10^{-18} g has major potential application for such clinical studies.[47]

8. Molecular Mechanisms of Aluminium-Induced Neurotoxicity

For any proof of the hypothesis that aluminium may be a significant risk factor in the aetiopathogensis of Alzheimer's disease it must be shown that it has neurotoxic actions capable of producing the hallmark neuropathological lesions of this disorder, especially β-amyloid deposition and neurofibrillary changes. Over 50 different neurotoxic effects of aluminium have been demonstrated[17] and these include actions on the cell nucleus, effects on enzymes and general metabolic processes, interference with cytoskeletal mechanisms, effects on membranes, and effects on neurotransmission. Most of these studies have not adequately taken into account the speciation of aluminium in solution and have used concentrations which would cause the precipitation of aluminium hydroxide on to cells and sub-cellular fractions incubated *in vitro*. It will be particularly important to investigate the effects of aluminium on cultured human cells using transferrin-mediated uptake to obtain intracellular concentrations similar to those in the ageing human brain. It seems increasingly likely that β-amyloid deposition may be central to the pathogenesis of Alzheimer's disease and a recent study has shown that a rare, early-onset, familial form of this disorder is associated with a genetic defect on chromosome 21 in which a single amino-acid substitution occurs in the transmembrane spanning region of the β-amyloid precursor protein, just outside of the COOH-terminal region of the β-amyloid peptide sequence.[48] Although the amino-acid substitution (Val → Ile) in this mutation is conservative, it has been pointed out that the base substitution in the mRNA for the APP may disrupt a regulatory stem loop containing a consensus base sequence (CACUGA) similar to that which forms part of the iron-responsive element involved in translational control of mRNA for ferritin and the transferrin receptor.[49] It is just possible that aluminium may influence translational control of APP through such a mechanism. Thus, studies on the effects of aluminium on control of APP gene expression (at both transcriptional and translational level) and also on the processing of APP are urgently required.

The tangle-like inclusions seen following the administration of aluminium to susceptible animal species such as the rabbit consist of abnormally phosphorylated, straight neurofilament proteins[50] which differ from the unique paired-helical filaments (PHFs) which are characteristic of neurofibrillary tangles in man.[51] There is remarkable species variation in the effects of aluminium on neurones and it is of interest that a component of the core protein sub-unit of PHFs isolated from Alzheimer brains consists of a hyperphosphorylated fragment of the microtubule binding protein, tau.[52, 53] Since the preferred intracellular binding site of aluminium is to vicinal phosphate groups[54] it will be of considerable importance to study the effects of aluminium on the phosphorylation of tau and other cytoskeletal proteins in human cells *in vitro*.

9. Conclusions

Although the cause of Alzheimer's disease is currently unknown, the possibility that aluminium may contribute to the aetiopathogenesis of this disorder cannot be readily dismissed. Aluminium is ubiquitous in the environment and appears to accumulate slowly in the central nervous system. The uptake of aluminium appears to be determined largely by the distribution of transferrin receptors which occur in greatest densities in regions of the brain that are selectively vulnerable in Alzheimer's disease. The precocious occurrence of β-amyloid deposition in renal dialysis patients exposed chronically to aluminium and the association of aluminium with plaques and tangles in Alzheimer's disease give further weight to a possible aetiopathogenic role. The «aluminium hypothesis» is also supported by several recent epidemiological studies. While these observations provide only circumstantial evidence for a causal relationship, together they provide a substantial case for further research in this field, particularly in relation to the molecular pathological effects of aluminium on human cells using transferrin-mediated uptake for loading.

A notable gap in the chain of evidence implicating aluminium in the aetiopathogenesis of Alzheimer's disease has been the lack of intervention studies in which the effects of limiting dietary intake of aluminium or of chelating agents on the progression of this disorder have been studied. If a recent report[55] that the chelating agent desferrioxamine slows the rate of cognitive decline by approximately 50% in patients with early Alzheimer's disease is confirmed, this will constitute a significant advance and open up exciting new prospects for therapeutic intervention.

Acknowledgements. The authors are indebted to the renal physicians and pathologists at the Royal Victoria Infirmary, Newcastle General Hospital and Freeman Hospital, Newcastle upon Tyne, Sunderland General Hospital and Ashington Hospital for their help with some of the studies reported here. Gifts of antibodies used in these investigations were received from K. Beyreuther, P. Davies, M. Landon and G. Perry. Mrs D. Hinds provided valuable secretarial help.

References

1. Henderson A.S.: *Acta Psychiat. Scand.*, **78**, 257-275, 1988.
2. Klatzo I., Wisniewski H.M., Streicher S.: *J. Neuropath. Exp. Neurol.*, **24**, 187-199, 1965.
3. Terry R.D., Peña C.: *J. Neuropath. Exp. Neurol.*, **24**, 200-210, 1965.
4. Driscoll C.T., Schecher W.D.: *Environ. Geochem. Health*, **12**, 28-49, 1980.
5. Ganrot P.O.: *Environ. Health Perspect.*, **65**, 363-441, 1986.
6. Epstein S.G.: *Environ. Geochem. Health*, **12**, 65-70, 1990.
7. Pullen R.G.L., Candy J.M., Morris C.M., Taylor G.A., Keith A.B., Edwardson J.A.: *J. Neurochem.*, **55**, 251-259, 1990.
8. McDermott J.R., Smith A.I., Wisniewski H.M.: *Neurology*, **28**, 809-814, 1979.
9. Morris C.M., Candy J.M., Oakley A.E., Taylor G.A., Mountfort S., Bishop H., Ward M.K., Bloxham C.A., Edwardson J.A.: *J. Neurol. Sci.*, **94**, 295-306, 1989.
10. Morris C.M., Candy J.M., Bloxham C.A., Edwardson J.A.: *Biochem. Soc. Trans.*, **18**, 647-648, 1990.
11. Burks J.F., Alfrey A.C., Huddlestone J., Norenburg M.D., Lewin E.: *Lancet*, **i**, 764-768, 1976.
12. Winkelman M.D., Ricanati E.S.: *Human Pathol.*, **17**, 823-833, 1986.
13. Brun A., Dictor M.: *Acta Microbiol. Scand. Sect A*, **89**, 193-198, 1981.
14. Candy J.M., McArthur F.K., Oakely A.E., Taylor G.A., Chen C., Mountfort S.A., Thompson J.E., Chalker P.R., Bishop H.E., Beyreuther K., Perry G., Ward M.K., Martyn C.N., Edwardson J.A.: *J. Neurol. Sci.* (submitted).
15. Gibson P.H.: *Neuropath. Appl. Neurobiol.*, **9**, 379-389, 1983.
16. Davies L., Wolska B., Hibich C., Multhaup G., Martin R., Simms G., Beyreuther K., Masters C.L.: *Neurology*, **38**, 1688-1693, 1988.
17. Crapper McLachlan D.R., Lukiw W.J., Kruck T.P.A.: *Environ. Geochem. Helath*, **12**, 103-114, 1990.
18. Sisodia S.S., Koo E.H., Beyreuther K., Unterbeck A., Price D.L.: *Science*, **248**, 492-495, 1990.
19. Candy J.M., Edwardson J.A., Klinowski J., Oakley A.E., Perry E.K., Perry R.H.: In *Senile Dementia of Alzheimer Type* (eds. Gispen W.H., Traber J.) Springer, Heidelberg, pp. 183-197, 1985.
20. Candy J.M., Oakley A.E., Klinowski J., Carpenter T.A., Perry R.H., Atack J.R., Perry E.K., Blessed G., Fairbairn A., Edwardson J.A.: *Lancet*, **i**, 354-357, 1986.
21. Edwardson J.A., Klinowski J., Oakley A.E., Perry R.H., Candy J.M.: In *Ciba Foundation Symposium, Vol. 121*, Wiley, Chichester UK, pp. 160-179, 1986.
22. Candy J., Oakley A., Gauvreau D., Chalker P., Bishop H., Moon D., Staines G., Edwardson J.: *Interdiscip. Topics Geront.*, **25**, 140-155, 1988.
23. Perry E.K., Oakley A.E., Candy J.M., Perry R.H.: *Neurosci. Lett.*, **25**, 321-325, 1981.
24. Perl D., Brody A.: *Science*, **208**, 297-299, 1980.
25. Perl. D.: *Personal communication*, 1990.
26. Still C.N., Kelly P.: *Neurotoxicology*, **4**, 125-131, 1980.
27. Vogt T.: *Sosiale og okonomiske studier*, **61**, Central Bureau of Statistics, Oslo, Norway, 1986.
28. Flaten T.P.: *Environ. Geochem. Health*, **12**, 152-167, 1990.

29. Martyn C.N., Osmond C., Edwardson J.A., Barker D.J.P., Harris E.C., Lacey R.F.: *Lancet*, **i**, 59-62, 1989.
30. Michel P., Commenges D., Dartigues J.F., Gagnon M.: *Neurobiol. Aging*, **11**, 264, 1990.
31. Birchall J.D., Chappell J.S.: *Lancet*, **i**, 953, 1989.
32. Flaten T.P., Odegård M.: *Food Chem. Toxic.*, **26**, 959, 1988.
33. French P., Gardener M.J., Dunn A.M.: *Food Chem. Toxic.*, **27**, 495-496, 1989.
34. Heyman A., Wilkinson W.E., Stafford J.A., Helms M.J., Sigmon A.H., Weinberg T.: *Ann. Neurol.*, **15**, 335-341, 1984.
35. Amaducci L.A., Fratiglioni L., Rocca W.A. et al.: *Neurology*, **36**, 922-931, 1986.
36. Graves A.B., White E., Koepsell T.D., Reifler B.V., Van Belle B., Larson E.B.: *J. Clin. Epidemiol.*, **43**, 35-44, 1990.
37. Broe G.A., Henderson A.S., Creasey H., McCusker E., Korten A.E., Jorm A.F., Langley W., Anthony J.C.: *Neurology*, **40**, 1698-1707, 1990.
38. McLaughlin A.I.G., Kazantzis G., King E., Teave D., Porter R.J., Owen R.: *Brit. J. Ind. Med.*, **19**, 253-263, 1962.
39. Rifat S.L., Eastwood M.R., Crapper McLachlan D.R., Corey P.N.: *Lancet*, **336**, 1162-1165, 1990.
40. Skalsky H.L., Carchman R.A.: *J. Amer. Coll. Toxicol.*, **2**, 405-423, 1983.
41. Taylor G.A., Vann R., Edwardson J.A.: Submitted for publication.
42. Wisniewski K.E., French J.H., Rosen J.F., Kozlowski P.B., Tenner M., Wisniewski H.M.: *Ann. New York Acad. Sci.*, **396**, 179-189, 1982.
43. Takashima S., Becker L.E.: *J. Neurol. Neurosurg. Psychiat.*, **48**, 61-64, 1985.
44. Garrett P., Spencer S., Muleahy D., Hanly P., O'Hare J.A., Murraghan D.J., Hanson J.S., Donohoe J.F., Carmody M., O'Dwyer W.F.: *Proc. EDTA-ERA*, **22**, 360-362, 1985.
45. Ferrier I.N., Leake A., Taylor G.A., McKeith I.G., Fairbairn A.F., Robinson C.J., Francis R.M., Edwardson J.A.: *Age and Ageing*, **19**, 368-375, 1990.
46. Farrar G., Altmann P., Welch S., Wychrij O., Ghose B., Lejeune J., Corbett J., Prasher V., Blain J.A.: *Lancet*, **335**, 747-750, 1990.
47. Baker J., Day J.P., Aitken T.W., Charlesworth T.R., Cunningham R.C., Drumm P.V., Lilley J.S., Newton G.W.A., Smithson M.J.: *Nucl. Instr. and Meth.*, **B52**, 540-543, 1990.
48. Goate A., Chartier-Harlin M.C., Mullan M., Brown J. et al.: *Nature*, **349**, 704-706, 1991.
49. Tanzi R.E., Hyman B.T.: *Nature*, **350**, 564, 1991.
50. Troncoso J., Sternberger L.A., Sternberger N.H., Hoffman P., Price D.: *Brain Research*, **364**, 295-300, 1986.
51. Kidd M.: *Brain*, **87**, 307-320, 1964.
52. Wischik C.M., Novak N., Thogersen H.C., Edwards P.C., Runswick M.J., Jakes R., Walker J.E., Milstein C., Roth M., King A.: *Proc. Natl. Acad. Sci. USA*·, 4506-4510, 1988.
53. Lee V.M.Y., Balin B.J., Otvos L., Trojanowski J.Q.: *Science*, **251**, 675-678, 1991.
54. Birchall J.D., Chappell J.S.: *Lancet*, **ii**, 1008-1110, 1988.
55. Crapper McLachlan D.R., Dalton A.J., Kruck T.P.A., Bell M.Y., Smith W.L., Kalow W., Andrews D.F.: *Lancet*, **337**, 1304-1308, 1991.

Aluminum (III) Toxicity and Blood-Brain Barrier Permeability

P.F. Zatta, M. Nicolini, B. Corain

Introduction

Biosphere occupies the external part of litosphere, which is mainly build up with oxygen, silicon and aluminum. Living organisms are exposed to an aluminum rich environment, in that its numerous derivatives are ubiquitous in soil, water and air. Non-sandy soils are basically made up with a great variety of aluminosilicates,[1] the most typical ones being clays. The atmosphere contains everywhere a defined background amount of aluminum as very finely powdered aluminosilicates.

Natural waters do possess very low Al^{III} concentrations, in that the thermodynamically stable form of the metal around pH 7 is solid $Al(OH)_3$, which allows a total (i.e., analytical), metal concentration of ca 10^{-7} M. Artificial or natural ligands[2] in the surface water as well as acute ecotoxicological phenomena such as acid rain[3, 4] may increase the actual analytical concentration of Al^{III} to a substantial extent.

In spite of the relative abundance of Al^{III} in the biosphere, no useful biological function of this metal center has been yet discovered, so far.[5] On the contrary, Al^{III} is now recognized as a toxic metal center and several reviews are available on this subject.[6-9]

Apparently, living organisms are carefully and effectively protected from Al^{III} aggression and some data[10] seem to suggest that naturally occuring Si^{IV} species might play a sort of planetarian «anthidote effect» to the ecotoxic potency of natural Al^{III}.

Al^{III} gains access to human body *via* two well established and one proposed pathways: the gastrointestinal tract and the lung tissue,[11] and the nose, i.e. through the olfactive neurons.[12] It is generally accepted that only 1% out of the daily intaken Al^{III} enters the blood stream and that ca 90% of this amount is excreted by the renal function. Therefore, only ca 1‰ of Al^{III} is accumulated in the human body, which normally contains 20-40 mg of metal in the adult.[13]

Abnormal exposures to Al^{III} are experienced by uremic patients[14] and by occupationally exposed workers.[15] In both cases, neurotoxic effects have been recognized,[16] which can be effectively eliminated only upon prevention strategies.

Aluminum Experimental Toxicology

The first experimental toxicological studies were reported by Siem[17] and Döllken[18] and were based on administration of aqueous solutions of aluminum salts. Neuronal degeneration in CNS in rabbits, cats and dogs was observed after subcutaneous injection with solution of sodium-aluminum tartrate or lactate. The observed degeneration was particularly evident in the lower cranial nerves. Thirty-two years later Seibert and Wells[19] reported lesions in the CNS of one series of rabbits to which large amounts of various salts of Al^{III} had been administred as aqueous solutions. Scherp and Church[20] injected intracranially solutions of aluminum lactate and suplhate at pH 6-7 (no aluminum hydroxyde precipitation was observed) in rabbits, mice and monkeys. As the consequence of this, rabbits developed neurological symptoms after 7-10 days after treatment, such as diminution in muscle tone, ataxia and eventually prostration. Convulsions occured in the terminal stage. Animals died 3-8 days after the onset of symptoms. Histological analysis revealed widespread nerve cells changes, showing a typical degeneration in the inflammatory reaction. Mice and monkeys receiving intracranial aqueous aluminum lactate (10 μg and 2.1 mg of Al^{III} respectively) developed the same syndrome few weeks after treatment.

Renewal of interest in Al^{III} toxicity aroused with the findings published by Klatzo et al.,[21] who serendipitously discovered that intracerebrally injected Holt's adjuvant (mainly $AlPO_4$) in rabbits, induced convulsions with neurofibrillary degeneration (NFD). At the first evaluation, the Al^{III}-induced NFD appeared reminiscent of that observed in neurons of patients affected by Alzheimer's disease (AD). However, further thorough investigations of the ultrastructure of these experimentally degenerated neurofilaments revealed that they were similar *but not identical* to those observed in AD affected neurons and Wisniewski and associates began to clarify in 1966[22] the topography of the observed neurofibrillary degeneration. In particular the spinal cord and the brain stem neurons of adult rabbits resulted to be particularly affected by Al^{III} with the limbic cortex, which appeared to be the most metal sensitive region of the cerebrum.

Uemura and coll.[23,24] provided a better understanding of the mechanism of metal-induced neurofibrillary degeneration in that they showed that rabbits subcutaneously injected with aluminum tartrate solutions developed neurofibrillary tangles at first in the brain stem and in the spinal cord (as it had been observed by Wisniewski et al. in 1966),[22] and eventually also in the hippocampus. Histological examination of the degenerated neurofilaments revealed an accumulation of large swirls of tangles of 10 nm neurofilaments in the cytoplasm of neurons, again ultrastructurally different from those of AD.

«Aluminum powders» injected into cerebrospinal fluid of mature rabbits induced a slowly progressive encephalomyelopathy characterized by alteration of posture, by myoclonic jercks and by muscle weakness.[25] Aluminum (III) derivatives were found inside the basal cisterns and in subarachnoid space of the spinal cord and in the central canal. Histologically, all areas of the grey matter resulted to be susceptible to the toxic effects of Al^{III}. Tangles developed in the spinal cord, medulla and pons. Subsequently neurofibrillary degenerations involved all grey matter structures except the putamen, the nucleus caudatus and the amygdala.

Numerous other experiments showed Al^{III}-induced neurofibrillary degeneration in «aluminum sensitive» animals such as cats, ferrets, mature and immature rabbits, dogs. On the contrary other species appeared to be remarkably resistant as is the case of guinea pigs, hamster, monkeys and man.[26]

Neurofibrillary changes due to the action of solutions or suspension of Al^{III} compounds were also produced in cell cultures as in the case of dorsal root ganglia,[27] murine,[28] and human neuroblastoma cells.[29]

Aluminum Neurotoxicology: Clinical Observations

Measurable concentrations of aluminum in brain tissues of human embryos have been reported, which gradually increased during gestation and reached the highest level in the newborn.[30] Data supporting the evidence that aluminum may be an essential element with regard to brain tissue development were presented by observations in animals.[31-35] A general agreement has been reached concerning the accumulation of Al^{III} in brain tissue during normal aging.[36,37]

The first case of human intoxication with neurological manifestations such as memory loss, tremor, jerking movements, impaired coordination, was described by Spofforth.[38] This seminal paper represents the case report of a clear association between Al^{III} and neurological disorders.

A peculiar case of «generalized aluminosis» of unknown origin in a 27 years old man, who died ten years later, was reported by Lapresle et al.,[39] who described a progressive neurological and mental deterioration. Macroscopic examination showed a global dilatation of ventricles with thinning of corpus callosum and grey lesions similar to plaques in the white matter. Histological analysis of brain tissue revealed a patchy demyelination and the presence in the white matter of depots of Al^{III} and phosphorus compounds with dimensions ranging from 1 to 200 μm. It is most interesting that in this case of aluminosis, the subject had no history of aluminum exposure.

A case of extensive aluminum fibrosis of lungs in a man 49 years old has been reported by McLaughlin et al.[40], followed by many others. The subject had worked for more than 13 years in the ball-mill room of an aluminum powder factory. The subject revealed symptoms referable to the CNS. The metal content in the body tissues revealed to be twenty times higher in brain and hundred and twenty times higher in the liver with respect to the control, which was estimated to be 0.6 μg Al/g w/w.[41-43] The subject became forgetful with difficulty in speeching and exhibited jerking movements. *Post-mortem* examination revealed healthy cerebral vessels. The brain was oedematous with dilatation of ventricles. Spinal cord was normal. Brain histological examination showed no evidence of abnormality.

A similar exposure-syndrome association was observed in other cases [44, 45] of subjects exposed to aluminum-containing dust. Neurological symptoms such as rapid progressive encephalopathy and cerebral degeneration, starting with focal clonic seizures in the legs and in the arms and aphasia with generalized convulsions, were associated with aluminum accumulation in brain.[40]

Welders and workers handling «aluminum powders» and aluminum sulphate appeared to have higher aluminum concentrations in blood and urine than non-exposed controls.[46] Blood metal concentrations were found to be lower than those associated with aluminum encephalopathy in patients under dialytic treatment. Workers exposed to aluminum flake powders and then retired, showed that aluminum is retained and stored in several compartments of body and eliminated from these compartments at different rates over many years. Finally long-term aluminum exposed welders exhibited more neuropsychiatric symptoms compared to non exposed ones.[47, 16]

Table I. **Aluminum accumulation in certain human conditions.**

1.	Dialysis encephalopathy.
2.	Alzheimer's disease.
3.	Amiothrophic lateral sclerosis.
4.	Parkinsonism-dementia complex of Guam.
5.	Down syndrome with manifested Alzheimer's disease.
6.	Industrial exposure (aluminum ball-mill workers).
7.	Neurofibrillary degeneration adjacent to hamartoma.
8.	Striatonigral syndrome.
9.	Alcohol dementia with patchy demyelination.
10.	Senile plaques of Alzheimer's disease.
11.	Aged human brain.

Aluminum and Human Encephalopathies

Aluminum has been implicated during the last two decades as a potential neurotoxic factor in different brain conditions.[48] A list of human conditions found to be associated with elevated aluminum concentrations in the brain is reported in Table I.[49]

Dialysis Encephalopathy

The role played by Al^{III} in this pathology was first described by Alfrey et al. in 1976 [50] (see also Alfrey in this volume) and its pathogenic role is now generally accepted.[51, 52] In particular, high levels of Al^{III} in brain tissues of long-term dialysis subjects were determined in *post-mortem* specimens.[53] In spite of the ascertained etiological role of Al^{III} accumulation in dialysis encephalopathy, the biochemical bases of metal neurotoxicity are unknown and they are still matter of investigation. The metal seems to concentrate mainly in the grey matter of the brain, but, in contrast with other pathologies where aluminum appears to be involved, neurofibrillary tangles were only rarely found.[54] Remarkably, samples of brain from patients died of dialysis dementia (DD) revealed the presence of aluminum in high concentration mainly in lysosomes of brain cells with metal concentration about thousand times higher than in the controls.[55] Similar findings were observed in rats after i.p. injection with water-soluble aluminum salts.[56]

Crapper et al.[57] reported that aluminum content expressed per g of DNA in the brains from patients who died of dialysis encephalopathy, was lower than in controls, although whole tissue concentrations were elevated ten to fifteen times above the controls.

Alzheimer's Disease

McLahlan first hypothesized a potential role of Al^{III} in the pathogenesis of AD[58, 59] showing high aluminum (III) levels in neurons with NFD typical of AD. Since that time, focal aluminum concentrations have been reported to be present in the hallmark features of subjects affected by AD, i.e. neurofibrillary tangles (NFT) and senile plaques (SP).[60-62] However, owing to the publication of contradictory analytical observations evidences connecting the presence of Al^{III} to the etiology of AD are still tangential and matter of intense investigation.

Perl and Brody offered in 1980[63] the first evidence of *focal* metal concentration by using electron microscopy coupled with microprobe analysis of neurons from AD subjects. NFT were found to be characterized by far higher aluminum concentrations than those observed in adjacent not NFT-bearing neurons. However, other authors[64] could not reproduce the same findings.

Candy et al.[65] succeeded in determining by X-ray microprobe analysis, *focal* metal concentrations in the core of senile plaques from AD affected brains. Again other authors could not confirm this observation by using scanning proton microprobe analysis[66] and electron probe analysis[67] and it has been proposed[68] that this disagreement could be due to subtle technical problems mainly dealing with sampling features.

In AD affected neurons aluminum is believed to accumulate inside the nucleus, albeit not homogeneously. In fact, Al^{III} has been found to be mainly bound to heterochromatin fractions as proven by the observation that the mean value of Al^{III} concentration was found to be 700 and 2000 μg Al/g DNA for nuclei and heterochromatin respectively in control brains while, levels equal to 1400 and 3800 μg Al/g have been reported in AD brains.[57] Recently Senitz et al.[69] raised in evidence in AD subject on the aluminum accumulation in neurons, glial cells, senile plaques, primitive plaques and intracortical congophilic vessels.

Parkinsonism-Dementia of Guam and Amyothrophic Lateral Sclerosis

Focal concentrations of Al^{III} have been detected also in neurons of subjects affected by Parkinson's dementia complex of Guam (PD) and by amyotrophic lateral sclerosis (ALS) as reported by Perl et al.[70] In PD, neurofilaments tangles are similar to those observed in AD, while, on the contrary, senile plaques are not observed. Ca^{II} and Al^{III} were distributed in cell bodies and in axonal processes of NFT-bearing neurons,[71] while no significant differences were noted in the content of magnesium, silicon, manganese and iron as shown by X-ray emission analysis.[70] Both elements depots occur within the same NFT-bearing hippocampal neurons. On the basis of the above, authors hypothesized that these elements are involved in NFT formation, suggesting a secondary hyperparathyroidism caused by abnormally low levels of environmental Ca^{II} and Mg^{II} in soil and water. Guam ALS subjects present cortical neurons with tangles of neurofilaments in the perikaryon similar to those found in Al^{III}-induced encephalopathy and not PHF as in AD (Table II).

Aluminum and Blood-Brain Barrier

Blood-brain barrier (BBB) is a multiple structure situated at the cerebral endothelium, the choroid plexuses and the outer arachnoid membranes restraining the movement of molecules and ions.[72] The occurence of modification of BBB and its relevance to AD is still a matter of controversy.[73-77]

It has been reported that in normal dogs,

Table II. **Comparative histopathological findings in some neurological diseases.**

Parkinson's Dementia	Amyotrophic Lateral Sclerosis	Creutzfeldt Jacob
- PHF similar to those in AD. - No SP were observed. - Al^{III} and Ca^{II} in cell bodies, in axonal processes of NFT bearing neurons.	- Tangles and neurofilaments similar to those found in A^{III}-induced encephalopathy and *not the PHF of AD.*	Degenerative neurological disorder without abnormal Al^{III} accumulation.

rats and in humans with renal failure, oral aluminum loading with $Al(OH)_3$, as phosphate binder, produces a significant increase of metal concentration in the brain. In patients with renal failure but not affected by *dementia dialitica* and even not yet under dialysis treatment, brain aluminum was found to be seven times more concentrated that in normal subjects. In patients affected by dialysis dementia aluminum was more than fifteen times more concentrated than in normal subjectes.[78] Apparently metal concentration in the brain becomes elevated as a consequence of metal loading and, or renal failure.

Brain aluminum was found to be also elevated in patients affected by other pathologies such as hepatic coma and metastate cancer.[78] Aluminum was found to be also elevated in a patient affected by sclerosing angioma in the frontal cortex able to produce a local pressure[79] by an alteration of BBB permeability.

The observation of significant and consistent increase of aluminum in the heterochromatin of neurons from AD affected patients was connected to a defective BBB. Thus, in authors opinion Al^{III} would be toxic not only owing to its ability of interaction with heterochromatin but also to its capability of damaging BBB and making it less impermeable to Al^{III} itself.

Banks and Kastin[80] reported that 100 mg/kg of Al^{III} injected as $AlCl_3$ solutions in adult rats increased of 45% the permeability of BBB to intracarotid injected ^{125}I-N-Tyr-DSIP with respect to the controls. The lack of correlation between changes in BBB permeability and decreased level of phosphorus in the blood suggested that the effect of Al^{III} should not be mediated by its effect on phosphorus metabolism. A similar effect was observed for the immunoactive δ-sleep-inducing peptide (DSIP), which is a light cycle dependent one. In fact, its ability of crossing the BBB in rats was found to significantly increase after treatment with aqueous $AlCl_3$.[81]

Wisniewski et al.,[82] reported that i.v. injected Al^{III} in rabbits as aqueous $AlCl_3$ accumulates in neurons one hour after injection. An attempt to evaluate the ability of Al^{III} in altering the permeability of BBB, thus facilitating its uptake by the animal brain, was carried out by Kim and Wisniewski[83] in rats. These researchers utilized intraperitoneal injections of aqueous solutions of $AlCl_3$ and aluminum lactate $Al(Lact)_3$ as well as of suspensions of $Al(OH)_3$.[83] ^{14}C-sucrose was used as tracer for monitoring Al^{III}-induced BBB permeability defects. The capillary permeability surface area to ^{14}C-sucrose was found to be elevated in animals treated with $AlCl_3$ and $Al(Lact)_3$, but not in those treated with $Al(OH)_3$. Remarkably, the permeability resulted to be transiently modified, and the conclusion was that high concentrations of Al^{III} alter the function of BBB in rats *in a reversible fashion.*

In a related study,[84] subcutaneously injected aqueous $AlCl_3$ (3.78 mg/d per 28 days), exhibited no toxicity in the rabbit. No significant differences were found in brain grey matter between the control and treated animals. As the brain tissue of animal treated had the lowest level of Al^{III} with respect to other tissues and organs, authors claimed that the permeability of BBB was only partially modified.

Aluminum levels were measured in the CSF of seventeen patients with infarction, eighteen with hemorrhage and sixteen with other acute ipoxy-ischemic insults or BBB damage. Al^{III} levels were significantly lower in brain hemorrhage than in brain infarction.[85]

Banks et al.[86] in treated mice, either i.p. or i.v., with aqueous $AlCl_3$ (up to 100 mg/kg) showed that the metal rapidly and profoundly inhibited the saturable system able to transport N-tyrosinated peptides Tyr-MIF-1 and the enkephalin from brain into the blood by acting on the basis of a noncompetitive inhibition. While the disappearance of Tc-pertechnetate-albumin or D-tyr-MIF-1 was not affected by Al^{III}, authors suggested that the metal center has to exert a highly selective effect on specific transport systems of BBB.

The membrane permeability of the primary cultures of brain microvessel endothelial cells monolayers can be modified by aqueous $AlCl_3$ at μmolar concentrations.[87] Al^{III} was found to reduce the pinocytosis with no effect on membrane lipid order. Authors suggested that the potential site of Al^{III}-induced alteration of BBB permeability originates at the luminal site of cells.

Glick hypothesized[88] that an altered form of albumin contributes to the pathogenesis of AD by having a greater affinity for Al^{III}. This altered protein should be able to cross the BBB more easily than the normal one and to compete with the normal protein in binding to brain neurons. Binding of this modified protein to neurons would facilitate Al^{III} uptake and in the same time impede Mg^{II} uptake.

In addition, ^{67}Ga bound to plasma transferrin was found to enter the brain giving raise to a marked regional distribution very similar to that observed for ^{125}I-transferrin.[89]

A schematic presentation of the involvement of aluminum (III) accumulation in different human conditions is depicted in Scheme 1.

Potential Al^{III} Carriers

Citrate and Low Molecular Weight Ligands

Aluminum (III) occurs in normal human plasma at concentrations in the 0.1-0.3 μM range.[90] Citric acid occurs in blood plasma at a concentration of about 0.1 mM.[91] Citrate binding of Al^{III} in aqueous solution has been des-

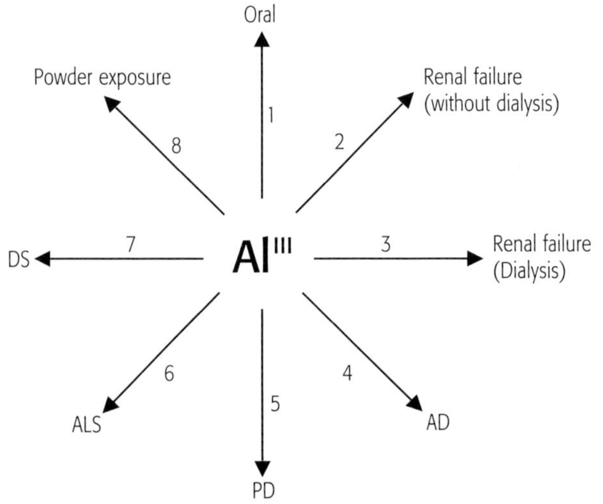

1. No particular accumulation in the brain of healthy subjects.
2. Moderate increase.
3. Al concentrates in neurons lysosomes.
4. Al accumulation in certain PHF-bearing neurons and in the core of SP.
5. Moderate increase in PHF.
6. Al in neurofilaments tangles.
7. Abnormal Al levels in neurons.
8. Al accumulation in brain and in other organs and tissues.

Scheme 1.

cribed in several studies,[90,92,93] and this carboxylate ligand is considered a major carrier of Al^{III} in mammalians plasma.[90] Thus, citrate is known to enhance aluminum solubility at the acidic pH of the stomach and consistently with this observation, after *in vivo* gavage of aluminum citrate in rats, plasma aluminum concentration resulted in peak plasma metal concentration within 40 minutes. On the contrary aluminum chloride gave a very small broad response in plasma aluminum (limited to ca 2%) with respect to citrate.[94]

In spite of this paramount relevance of citrate in Al^{III} uptake, the enhancement mechanism is not simple and not easy to be rationalized. In fact, in model experiments it has been demonstrated that the lipid bilayer permeation by the neutral aluminum-citrate complex is remarkably slow ($P \cong 1 \times 10^{-11}$ cm/sec).[95] Therefore, the observed increases of Al^{III} levels in the blood and in the tissues of experimental animals[96] and man[97] cannot be explained by the involvement of the[90] neutral aluminum-citrate complex, which dominates at the stomach acidity, in metal permeation of the naturally occuring gastric membranes.

Albumin

Albumin is present in the plasma at a concentration of 40 g/L.[98] Although albumin is a carrier of Al^{III} in the plasma[99] it is much too weak as a metal ion binder at physiological pH values[90] to be able to effectively compete with other much stronger physiological Al^{III} carriers, such as citrate and transferrin (Tf).[90] However, thanks to the high albumin blood concentration unspecific Al^{III} binding can occur and since albumin has been found in the senile plaques[75] from AD subjects, its possible involvement as an unspecific carrier of Al^{III} through BBB into the brain should not be neglected, at least in principle. The physicochemical properties of albumin can be changed by glycosylation and cationization and this can influence the vascular permeability of the protein *in vivo* and *in vitro*.[100,101] The modified protein appeared to have a high affinity for macrophages[102] and also it could be ingested in microcapillaries preparation from rat epididymal fat.[101] Glycosylated albumin exhibits a higher pI resulting in a higher affinity for the brain microvessels.[103]

Transferrin

Aluminum coordination by transferrin (Tf) can be depicted by the following equations:

1. $Tf + Al^{3+} + HCO_3^- \rightleftarrows Tf\text{-}Al^{3+}\text{-}HCO_3^-$

2. $Tf\text{-}Al^{3+}\text{-}HCO_3^- + Al^{3+} + HCO_3^- \rightleftarrows$
   $\rightleftarrows Tf\text{-}Al_2^{6+}\text{-}\ (HCO_3)_2^{2-}$

The glycoprotein transferrin is recognized as the major plasma iron transport protein.[104,105]

Human Tf binds 2 mol of Al^{III}/mol of protein. Since Tf is only 30% saturated with Fe^{III},[106] it can be regarded as a potential ligand for Al^{III} under physiological conditions, without interfering with the sites occupied by iron (K'_{Al} much lower compared to K'_{Fe}).[107] However, it has been proposed that aluminum, when bound to Tf, inhibits iron uptake partly by down-regulating transferrin receptor (RTf) expression and also interferring with intracellular release of iron from Tf.[108] It is likelihood that Al^{III} as well as other toxic metals which bind tightly to Tf may enter the brain by this pathway. Remarkably however, many brain areas with high RTf density appear unrelated to the iron-transport protein,[109] thus suggesting other physiological role for Tf.

Tf has been also identified in oligodendrocytes[110] that are responsible for the synthesis and the maintenance of myelin in the CNS. RTf's have been identified on the endothelium of brain capillaries.[111] ^{125}I-Tf perfused in rats resulted in a receptor-mediated uptake of Tf into the endothelium of the BBB.[112]

In rats injected with aqueous $AlCl_3$[113] the metal in plasma was linearly related to Al^{III} in blood. Gel filtration of plasma from Al^{III}-treated animals showed that a large part of Al^{III} resulted to be bound to high molecular weight proteins (HMWP) such as albumin and Tf. However, a significant fraction of Al^{III} was found to be bound to other non-protein macromolecules of about 10 KD.[113] In contrast with this, other authors identified albumin and Tf as the only Al^{III} macromolecular carriers.[114] Other studies could not identify serum Tf as a major binding protein.[115,116] According to Trapp[114] these controversial findings can be explained on the basis of experimental artifacts, for instance due to inappropriate concentrations of HCO_3^- in the chromatographic column buffer.

On the basis of a thorough investigation,[117,118] Al^{III} was shown to bind to several carriers i.e., HMWP, albumin, low molecular weight proteins (LMWP), polypeptides and inorganic anions, in the serum from patients on long-term dialysis. Thus serum aluminum concentration in dialysis patients was reported to be distributed among five major peaks in relevant chromatographic preparations. Four peaks were associated with proteins and one was attributed to a LMW species, being this one a plausible candidate as effective carrier for crossing BBB.

A clinical implication of abnormal levels of plasma Tf was proposed by Farrar et al..[119] They observed that plasma Tf in AD patients was lower than in the controls and concomitantly, a higher loading of Tf-Fe^{III} was evident. Consequently authors speculated that the deficiency of Al^{III}-binding sites would leave more «unbound» Al^{III}, which could move readly into the brain, where it can express its neurotoxicity. Although we believe that this clinical observation may be worthwhile, the existence of «unbound» Al^{III} in the blood is unrealistic on the basis of thermodynamic observations.[90]

A new facet of the role played by Tf as a major carrier of Al^{III} through the BBB might appear by looking at Tf glycosylation. The presence of 2 glycans in human Tf has been established more than a decade ago[120] as structured in a biantennary N-acetyllactosamine type.[121] However, a large glycannic variant has been reported for this protein and the functions of Tf glycans are not yet fully defined. Furthermore, heterogeneity in the distribution of glycannic variant of Tf in pathological human serum has been largely demonstrated.[122] Although it has been reported that the glycosylation of Tf is not necessary for specific binding to its receptor, the affinity of the binding can be influenced by the presence or absence of carbohydrate residues.[123] At the present time there are no indications that Fe^{III} or Al^{III} transport into the cell is influenced by a modified glycosylation of Tf and there are no information on the modification of Tf glycosylation in AD. However, preliminary experiments carried out in these laboratories showed that tris-(acetylacetonate)aluminum(III) does alter the activity of β1,4 galactosyltransferase in human serum. Altered glycosylation in AD features has been reported by Vorbrodt et al.[124]

The Interaction of Al^{III} Artificial Toxins with Biological Systems: A Heuristic Model

In pure phenomenological terms, the interaction of Al^{III} with (and the revelance to) biological systems is documented by an impressive number of papers.[125] The toxicological literature is particularly rich[26] and a considerable body of information is also available on the interaction of «Al^{III} solutions» with enzymatic systems.[11]

In spite of this flowrishing literature, direct information on the molecular bases of aluminum biological activity are incredibly scanty.[125]

The reason for this unusual situation is at least twofold, i.e. i) in the very large majority of the experimental toxicology protocols Al^{III} was administrered either under ill-defined pH conditions (common case) or as ill-defined mixture of various chemical species in the neutrality range[126] and ii) very few research groups investigated on the specific biological activity of Al^{III} administered under precise chemical forms and on their reactivity with small biomolecules or with biologically relevant macromolecules.

As a matter or fact, researchers are now challenged to interprete an enormous amount of biological, Al^{III}-related, phenomena on the basis of a dearth of bioinorganic molecular information.

A conceptual experimental breakthrough aimed to circumvent points i) and ii) was indipendently proposed in these and other laboratories in 1986.[127,128]

The basic intuition is the employment of Al^{III} complexes bearing:

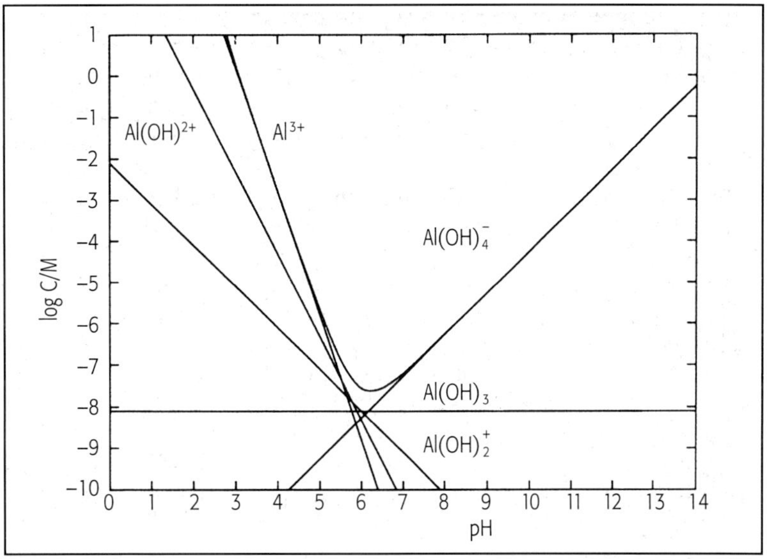

Fig. 1. Calculated solubility and speciation diagram for Al^{III} in water in the absence of ligands different from H_2O and OH^-. In the formulas metal-coordinated water molecules are omitted. (Data from ref. 126).

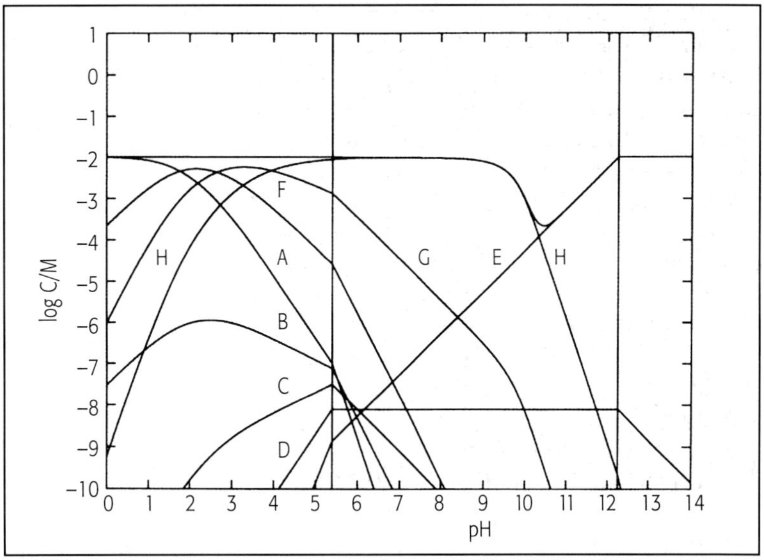

Fig. 2. Solubility and speciation of Al^{III} as a function of pH in the presence of maltolate 0.03 M; log K_s= $=10^{22.5}$, $t=25°C$. A: Al^{3+}; B: $Al(OH)^{2+}$; C: $Al(OH)_2^+$; D: $Al(OH)_3$; E: $Al(OH)_4^-$; F: $Al(malt)^{2+}$; G: $Al(malt)_2^+$; H: $Al(malt)_3$. (Data from ref. 126)

a) a defined molecular structure in solutions under the actual employment conditions;
b) net zero electrical charge;
c) high hydrolitic stability;
d) defined hydrophilic-lipophilic character measured as n-octanol/water partition coefficients.[129, 130]

The combination of these requisites offers the possibility of using artificial toxins for which it is possible to calculate (and ideed potentially to experimentally check) the metal speciation when the biological system is exposed to the synthetic toxin. Thus, under these conditions, dose-response relationships become reliable and molecular models of Al^{III} toxicity may be reliably developed.

In the absence of strongly coordinating agents, the aqueous solution chemistry of Al^{III} is seemingly simple (Fig. 1). The thermodynamically predicted species which dominate at

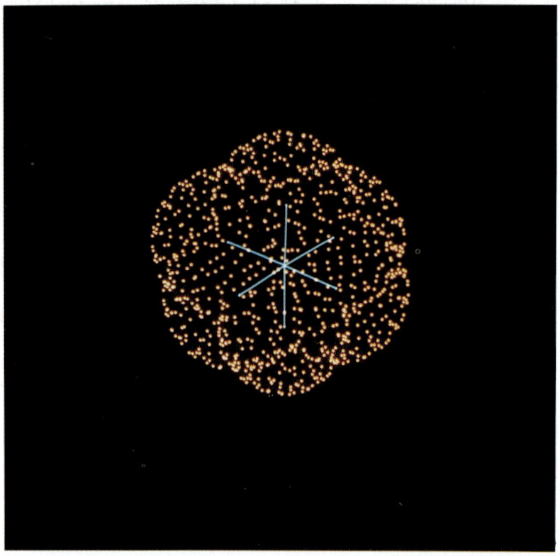
A

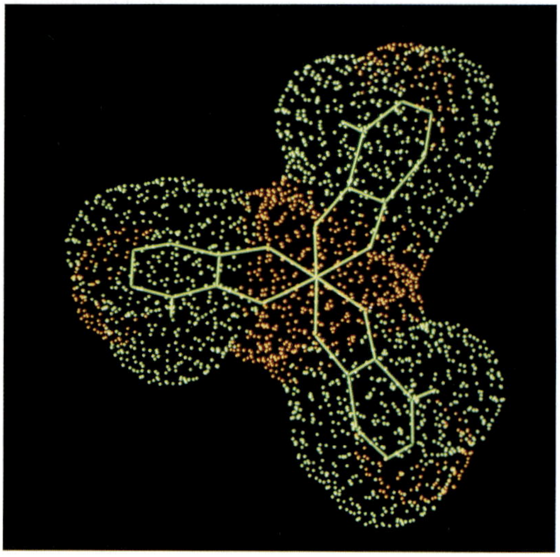
B

Fig. 3. Pictorial presentation (computer graphics) of the molecular structures of A) $[Al(OH_2)_6]^{3+}$; B) $Al(malt)_3$; C) $Al(acac)_3$. The reddish areas correspond to the oxygen atoms, i.e. the hydrophilic parts of coordination sphere. The greenish areas correspond to the hydrophobic parts. (Courtesy of R. Pavan and M. Tognolin, Dept. of Organic Chemistry, University of Padova).

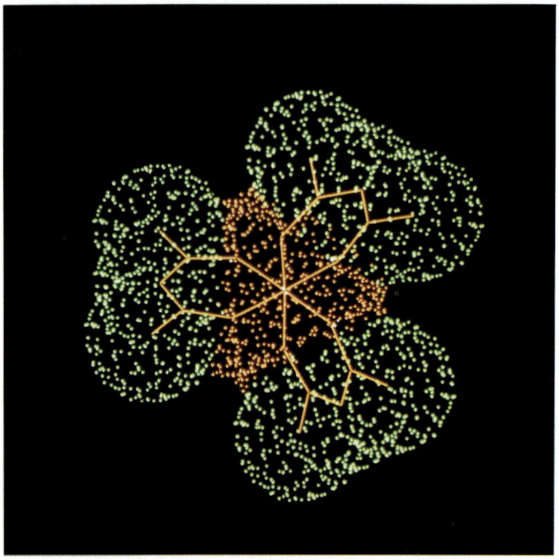
C

pH 7.5 should be $Al(OH)_4^-$ ($10^{-6.86}$ M) and « $Al(OH)_3$ » ($10^{-8.10}$ M) at an expected total metal concentration equal to ca 10^{-7} M.[126]

In fact, owing to the intimate inertness of the neutral Al-complexes to ligand displacements[90] around pH 7.0, the equilibration time needed to letting the oligomeric forms precursors of $Al(OH)_3$ to fully equilibrate with the solid amorphous phase, may vary from days to months.[131]

On the contrary, complexes such as $Al(acac)_3$ (acac=acetylacetonate) and $Al(malt)_3$ (malt=maltolate) appear to meet the requisites necessary to by-pass the drawbacks outlined in points i) and ii) (Fig. 2).

$Al(acac)_3$ is reasonably water soluble (7.8×10^3 M at ca 20 °C), albeit mainly lipophilic in character,[130] and it can survive thanks to kinetic reasons at analytical concentrations as low as 10^{-4} M, for which the thermodynamic calculations (based on eq. 4) predict extensive dissociation to give eventually Hacac and $Al(OH)_3$.[131] Quite similar considerations are valid for $Al(malt)_3$, which is definitely more soluble in water (ca 6×10^{-2} M) and hydrophilic in character.[130]

A pictorial presentation of the structures of $Al(acac)_3$, $Al(malt)_3$ and $[Al(OH_2)_6]^{3+}$ aimed to emphasize the difference in their lipophilic-hydrophilic character, as the consequence of distinct differences in the chemical nature of their coordination spheres, is presented in figure 3.

The equation dealing with the metal speciation in water of artificial toxins bearing bidentate monoionic ligands, are the followings:

Fig. 4. Calculated domains for the existence of the species $Al(malt)_3$ as a function of metal analytical concentration and of pH; t=25 °C. (Data from ref. 126)

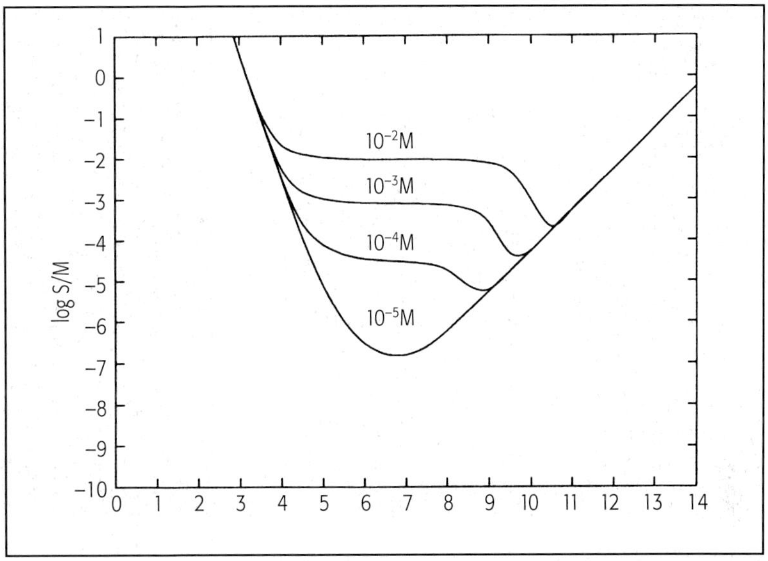

$$Al(L\text{-}L)_3 + 6H_2O \xrightleftharpoons{K_i} [Al(H_2O)_6]^{3+} + 3\ L\text{-}L^- \quad (1)$$

$$3\ L\text{-}L^- + 3\ H_2O \xrightleftharpoons{K_b} 3\ H(L\text{-}L) + 3\ OH^- \quad (2)$$

$$Al(H_2O)_6^{3+} + 3\ OH^- \xrightleftharpoons{K_s} Al(OH)_3 + 6\ H_2O \quad (3)$$

$$Al(L\text{-}L)_3 + 3\ H_2O \xrightleftharpoons{K_{dec}} Al(OH)_3 + 3H(L\text{-}L) \quad (4)$$

A proper value of K_i will determine the thermodynamically predicted speciation of Al^{III} at any pH value for any analytical concentration. The predominance of the species $Al(L\text{-}L)_3$ (i.e.$[Al(L\text{-}L)_3] / [Al^{III}]_{tot}$) will be facilitated by smaller K_i and higher $[Al^{III}]_{tot}$ values. This concept is illustrated in figure 4 for the case of $Al(malt)_3$.

It can be seen that a) the predicted stability «window» for $Al(malt)_3$ becomes narrower, and b) the $[Al(malt)_3] / [Al^{III}]_{tot}$ ratio become lower and lower *with decreasing total metal concentration*. However, as noticed above, kinetic factors might become decisive for extending the actual survival of the neutral species to lower analytical concentrations. Thus, spectrophotometrically stability tests carried out in neutral water solutions for $Al(acac)_3$ and $Al(malt)_3$ [132] at 25 °C and 37 °C have revealed that the rate of hydrolysis of both species (eq. 4) is so slow that, e.g. ca 60% of $Al(malt)_3$ is still present in solutions after 100 h at 50 μM analytical concentration at 37 °C.

In conclusion, in these cases the combination of thermodynamic and kinetic factors, do preserve the designed metal speciation down to μM concentrations for remarkably long times.

Kinetic factors are of paramount importance also for the speciation of $Al(lact)_3$ at pH 7.5.[126] $Al(lact)_3$ undergoes nearly quantitative loss of the three ligands, originally coordinated to the metal center in the solid state,[133] after dissolution in water and pH adjustment to neutrality. Under these conditions metastable species, including $Al(OH)_3(H_2O)_3$ dominate the solution chemistry of this largely used toxin down at least ca 5×10^{-3} M analytical concentration.

The whole of the experimental toxicological (*in vivo* and *in vitro*) and preliminary enzymological experiments carried out in these laboratories [135] appear to strongly support the choice

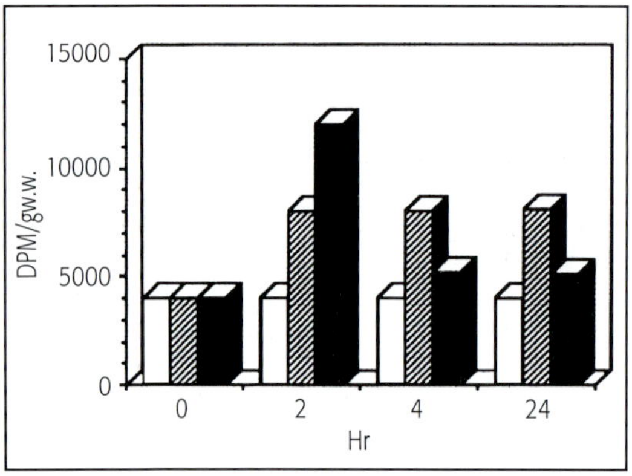

Fig. 5. Cerebrovascular permeability to ^{14}C-sucrose, in rats, 2, 4 and 24 h after injection of (□) NaCl; (▨) $Al(acac)_3$: 2.2mg Al/Kg body weight; (■) $Al(malt)_3$: 2.2 mg Al/Kg body weight.

of our heuristic model based on the utilization of diversely stable and lipophilic-hydrophilic synthetic toxins.[136] Thus, in toxicological experiments on rabbits, erythrocytes and enzymological work,[135] large and consistent speciation effects have been observed with reliable dose-response relationship. Analogous speciation-related observations, albeit dealing with a more limited scope, have been recently published by other groups.[137-140]

As noted above, i.v. injection of *acidic* aqueous solutions of Al^{III} in rats produces a noticeable and reversible increase of BBB permeability to ^{14}C-sucrose.[88] Although this sort of «picklock» effect has to be undoubtedly ascribed to the action of the metal on BBB, the administration protocol represents a typical example of completely uncontrolled speciation conditions. Thus, the injection of $[Al(H_2O)_6]^{3+}$ into the rat blood stream will be immediately followed by a mess of chemical events i.e. local precipitaion of $Al(OH)_3$ and, or reaction of Al^{III} with citrate, Tf, albumin, etc.

In order to confirm (and possibly to extend the relevance of) this experiment, we recently investigated on the effect of Al^{III} on BBB permeability by administering the metal under controlled speciation conditions.[141] Thus, we utilized $Al(acac)_3$ (hydrolytically stable and lipophilic) and $Al(malt)_3$ (hydrolitically stable and hydrophilic). As expected, while the action of Al^{III} on BBB was confirmed, new aspects of this biological effect became apparent. While $Al(malt)_3$ produced a transient (i.e. reversible) effect lasting few hours, $Al(acac)_3$ caused an almost irreversible increase of BBB permeability. In fact, the return of the permeability to normal values required more than 24 h. Some relevant data are depicted in figure 5.

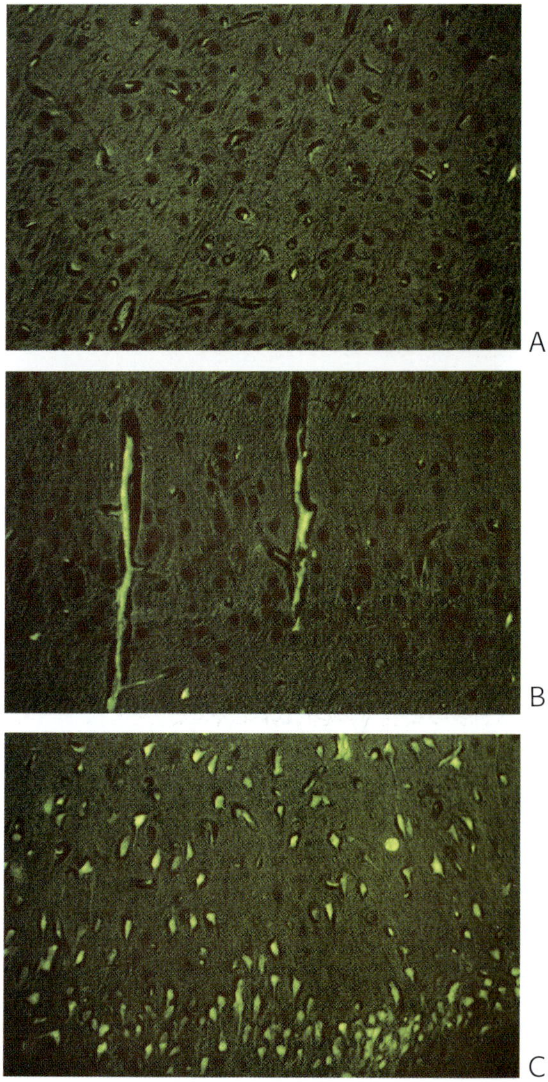

Fig. 6. Morin staining: Cerebral cortex field from rats after 24 h from treatment with A) NaCl; B) $Al(malt)_3$: 2.2 mg Al/Kg; C) $Al(acac)_3$: 2.2 mg Al/Kg. Magnification: × 350.

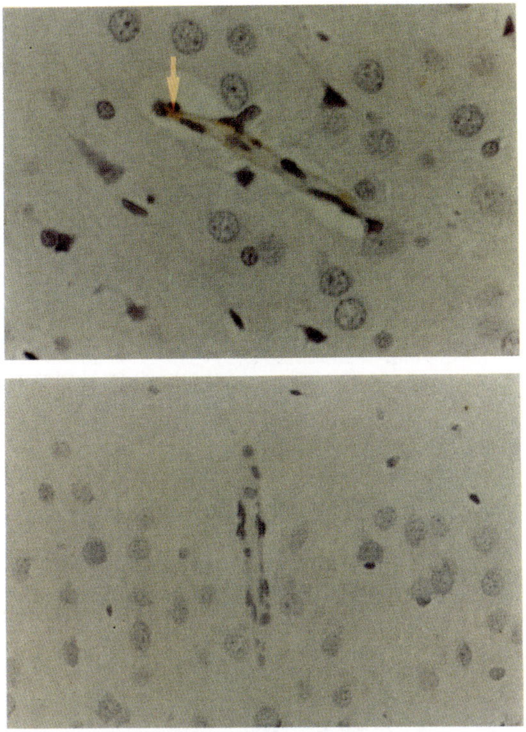
A
B

Fig. 7. Rat cerebral cortex from animal treated with A) $Al(acac)_3$: 2.2 mg Al/Kg body weight. Microcapillaries react positively with BS-I Lectin. B) Control. Magnification: × 400.

Aluminum accumulation could be observed only in the neurons of the cerebral cortex from rats treated with the more lipophilic toxin $Al(acac)_3$ (Fig. 6). On the contrary, and remarkably, the hydrophilic species $Al(malt)_3$ gave positive morin staining only in the microvessels.

Another marked speciation effect was observed in connection with the glycosylation of the proteins of the luminal side of the microcapillaries. The lectin *Bandeiraea simplicifolia* (BS-I) is able to recognize non-reducing GalNAc residues, thus offering the possibility to localize such a sugar residue in cellular and even in subcellular compartments.[142] Microcapillaries from animals treated with $Al(acac)_3$ can be recognized by BS-I lectin while the same lectin does not interact with the microcapillaries from animals treated with $Al(malt)_3$, $Al(lact)_3$ and the controls.

These results apparently reveal that $Al(acac)_3$ and not $Al(malt)_3$, (data not shown) interferes with the process of glycosylation of the luminal side of microcapillaries. The change in the distribution of endothelial surface glycoconjugates associated with alteration of BBB permeability have been documented for brain microvessels by other authors.[143]

Acknowledgements. This work was partially supported by Parke Davis and Fidia Research Laboratories, Italy. Authors wish to thank dr. M. Favarato for partial collection of literature data.

References

1. Ferguson J.E.: in *Inorganic Chemistry and the Health.* Pergamon Press, 1982.
2. Stevenson F.J., Vance G.F.: in *The Environmental Chemistry of Aluminum.* Sposito G. (Ed.), CRC Press Inc., Boca Raton, Florida, 1989.
3. Driscoll S.C.T., Naker J.P.B., Bisogni J.J., Schonfield C.: *Nature*, **284**, 161, 1980.
4. Schindler D.W.: *Science*, **239**, 149, 1988.
5. MacDonald T.L., Martin G.: *TIBS*, **13**, 15, 1988.
6. Wills M.R., Savory J.: *Lancet*, **1**, 29, 1983.
7. Boegman R.J., Bates L.A.: *Can. J. Physiol. Pharm*, **62**, 1010, 1984.
8. Elinder C.-G., Sjogren B.: in *Handbook of the Toxicology of Metals.* 2nd Ed., Friberg J., Nordberg G.F., Vonk V. (Eds), Elsevier, North Holland, 1986, pg. 1.
9. Krishnan S.S., MacLachlan D.R., Dalton A.J., Krishnan B., Fenton S.S.A., Harrison J.E., Kruck T.: in *Essential and Toxic Elements in Human Health and Disease.* Alan Liss Inc., 1988, pg. 645.
10. Birchall J.D., Exley C., Chapell J.S., Philips M.J.: *Nature*, **338**, 146, 1989.
11. Ganrot P.O.: *Environm. Health Perspect.*, **65**, 363, 1986.
12. Roberts E.: *Neurobiol. of Aging*, **7**, 587, 1986.
13. Pennington J.A.T., Jones J.W.: in *Aluminum and Health.* Gitelman H.J. (Ed), Marcel Dekker Inc., New York, 1989.
14. Wills M.R., Savory J.: in *Metal Ions in Biologica Systems.* Sigel H., Sigel A. (Eds), Marcel Dekker Inc., New York, vol. 24, pg. 315.
15. Ljunggren K.G., Lidums V., Sjogren B.: *Brit. J. Ind. Med.*, **48**, 106, 1991.
16. Rifat S.L., Eastwood M.E., Crapper McLachlan D.R., Corey P.N.: *Lancet*, **1**, 1162, 1990.
17. Siem. Inaug. Dorpat, 1886.
18. Döllken P.: *Arch. Exp. Path. u. Pharm.*, **40**, 98, 1987.

19. Seibert F.B., Wells H.G.: *Arch. of Pathol.*, **8**, 230, 1929.
20. Scherp H.W., Church F.: *Proc. Soc. Exp. Biol.*, **36**, 851, 1937.
21. Klatzo I., Wisniewski H.M., Streicher E.: *J. Neuropath. Exp. Neurol.*, **24**, 187, 1965.
22. Wisniewski H.M., Narkiewicz O., Wisniewski K.: *Acta Neuropathol.*, **6**, 211, 1966.
23. Uemura E., Ireland W.P.: *Exp. Neurol.*, **1**, 9, 1984.
24. Uemura E., Ireland W.P.: *Exp. Neurol.*, **89**, 530, 1985.
25. Bugiani O., Ghetti B.: *Neurobiol. of Aging*, **3**, 209, 1982.
26. Sturman J.A., Wisniewski H.M.: in *Metal Neurotoxicity.* Bondy S.C., Prasad K.N. (Eds), CRC Press Inc., Boca Raton, Florida, 1988, pg. 61.
27. Seil F.J., Lampert P.W., Klatzo I.: *J. Neuropath. Exp. Neurol.*, **28**, 74, 1969.
28. Miller C.A., Levine E.M.: *J. Neurochem.*, **22**, 751, 1974.
29. Cole G.M., Wu K., Timiras P.S.: *Int. J. Dev. Neurosc.*, **3**, 23, 1985.
30. Leonov V.A.: *Vesti Akad. Navuk Belarus. SSR Ser. Biyal. Navuk.*, N. 1, **151**, 1, 1956.
31. Voinar A.O., Soroka V.R., Sabadash E.V.: *Mikroelem. Nerv. Sist. Baku*, **6**, 1963; 6, 1966.
32. Soroka V.R.: *Bull. Exp. Biol. Med. (USSR)*, **47**, 196, 1959.
33. Soroka V.R.: *Tr. Stalinskogo Med. Inst.*, **19**, 187, 1961.
34. Soroka V.R.: *Tr. Stalinskogo Med. Inst.*, **19**, 183, 1961.
35. Sabadash E.V., Soroka V.R.: *Fiziol. Zh. SSSR im I.M. Sechenova*, **48**, 994.
36. Crapper D.R., De Boni U.: *Can. Psychiatr., Ass. J.*, **23**, 229, 1978
37. Markesbury W.R., Ehmann W.D., Hossaim T.I.M., Alauddin M., Goodin D.T.: *Ann. Neurol.*, **10**, 511, 1981.
38. Spofforth J.: *Lancet*, **1**, 1301, 1921.
39. Lapresle J., Duckett S., Galle P., Cartier L.: *Compt. Rend. Soc. Biol.*, **2**, 282, 1975.
40. McLaughlin A.I.G., Kazantzis G., King E., Teare D., Porter R.J., Owen R.: *Brit. J. Ind. Med.*, **19**, 253, 1962.
41. Tripton I.H., Cook M.J., Steiner R.J., Foland J.M., McDaniel K.K., Fentress S.D.: *Oakridge Natl. Lab. Reports*: Central Files a) 57-2-3; b) 57-2-4; c) 57-11-33, 1957.
42. Tripton I.H., Cook M.J., Foland J.M., Rittner J., Harwick M., McDaniel K.K.: *Oakridge Natl. Lab. Reports*: Central Files N. 58-10-15, 1958.
43. Tripton I.H., Cook M.J., Foland J.M., Rittner J., Harwick M., McDaniel K.K.: *Oakridge Nath. Lab. Reports*: Central Files N. 59-8-106, 1959.
44. Goralewski G.: *Die Aluminumlunge* (Arbeitmedizin N. 26) Barth, Leipzig, 1950.
45. Nagelschmidt G.: *Brit. J. Ind. Med.*, **17**, 247, 1960.
46. Sjogren B., Lundberg I., Lidumus V.: *Brit. J. Ind. Med.*, **40**, 301, 1983.
47. Sjogren B., Gustavsson P., Hogsted C.: *Brit. J. Ind. Med..*, **47**, 704, 1990.
48. Crapper McLachlan D.R., Lukiw W.J., Kruck T.P.A.: *The Can. J. Neurol. Sciences.*, **16**, 490, 1989.
49. Crapper McLachan D.R., De Boni U.: in *Aluminum Neurotoxicity.* Alan Liss Inc. (Ed), Pathotox Publ. Inc., Park Forest South, Illinois, 1980, pg. 3.
50. Alfrey D.C., LeGendre G.R., Kaehny W.D.: *N. Engl. J. Med.*, **294**, 184, 1976.
51. Alfrey A.C.: *Clin. Nephrol.*, **24**, S15, 1985.
52. Arieff A.I.: *Am. J. Kidney Dis.*, **6**, 317, 1985.
53. McDermott J.R., Smith A.I., Word M.K., Parkinson I.S., Kerr D.N.S.: *Lancet*, **1**, 901, 1978.
54. Bugiani O., Brancaccio D., Ghetti B.: *Clin. Nephrol.*, **24**, S20, 1985.
55. Galle P., Chatel M., Berry I.P., Menault F.: *Nouv. Press Med.*, **8**, 4091, 1979.
56. Galle P., Berry J.P., Duckett S.: *Acta Neuropath.*, **40**, 245, 1980.
57. Crapper D.R., Quittkat S., Krishnan S.S., Dalton A.J., De Boni U.: *Acta Neuropath.*, **50**, 19, 1980.
58. Crapper D.R., Krishnan S.S., Dalton A.J.: *Science*, **180**, 54, 1973.
59. Crapper D.R., Dalton A.J.: *Physiol. Behav.*, **10**, 935, 1973.
60. Trapp, G.A., Miner G.D., Zimmerman R.L., Mastri A.R., Heston L.L.: *Biol. Psychiatr.*, **13**, 709, 1978.
61. De Boni U., Scott J.W., Crapper D.R.: *Histochem.*, **40**, 31, 1974.
62. Crapper D.R., Krishnan S.S., Quittkat S.: *Brain*, **99**, 67, 1976.
63. Perl D.P., Brody A.R.: *Science*, **208**, 297, 1980.
64. Kobayashi S., Hirota N., Saito K., Utsuyama M.: *Acta Neuropath.*, **4**, 47, 1987.
65. Candy J.M., Oakley A.E., Klinowski J., Carpenter T.A., Perry R.H., Atack E.J.R., Perry E.K., Blessed G., Fairbairn A., Edwardson J.A.: *Lancet*, **1**, 354, 1986.
66. Landsberg J.P., McDonald B., Roberts J.M., Grime G.M., Watt F.: *Neurobiol. of Aging*, **11**, 317, 1990 (abstract 266).
67. Chafi A.H., Hauw J.-J., Rancurel G., Berry J.-P., Galle C.: *Neurosci. Letters*, **123**, 61, 1991.
68. Edwarson J.A., Oakley A.E., Pullen R.G.L., McArthur F.K., Morris C.M. et al.: in *Aluminum in Food and the Environment.* Massey R.C., Taylor D. (Eds), The Royal Soc. of Chemistry, London, 1989, pg. 20.
69. Senitz D., Bluethner K.: *Zentralblatt f. Allgemeine Pathol. u. Pathol. Anatonie*, **136**, 329, 1990.
70. Perl D.P., Gajdusek D.C., Garruto R.M.,

YANAGIHARA R.T., GIBBS Cj.: *Science*, **217**, 1053, 1982.
71. GARRUTO R.M., FUKATSU R., YANAGIHARA R., GAJDUSEK D.C., HOOK G., FIORI C.E.: *Proc. Natl.Acad. Sci-USA*, **81**, 1875, 1984.
72. SEGAL M.B., ZLOKOVIC B.V.: in *The Blood-Brain Barrier, Amino Acids and Peptides*, Kluwer Academic Publ. Dordrecht, London, 1990, p. 1.
73. GLENNER G.G.: in *Molecular Biology and Genetics of Alzheimer's Disease*, MYATAKA YT., SELKOE D.J., IHARA Y. (Eds.), Excerpta Medica, Amsterdam, 1990, pg. 69.
74. SCHEIBEL A.B.: in *Senile Dementia: Outlook for the Future*, Alan Liss Inc., New York, 1984, pg. 137.
75. WISNIEWSKI H.M., KOZLOWSKI P.B.: *Annals of the New York Acad. Sci.*, **396**, 1982, pg. 119.
76. ROZENMULLER J.M., EIKELENBOOM P., KAMPHORST W., STAM F.C.: *Neurobiol. of Aging*, **9**, 383, 1988.
77. MIYAKAWA T., KURAMOTO R.: *Ann. Med.*, **21**, 99, 1989.
78. ARIEFF A.I., COOPER J.D., ARMSTRONG D., LAZARIWITZ V.C.: *Ann. Intern. Med.*, **90**, 741, 1979.
79. LISS L., EBNER K., COURI D.: *Human Pathol.*, **10**, 104, 1979.
80. BANKS W.A., KASTIN A.J.: *Psychopharmacol.*, **86**, 84, 1985.
81. BANKS W.S., KASTIN A.J., SELZNICK J.K.: *J. Neurosc. Res.*, **14**, 347, 1985.
82. WISNIEWSKI H.M., STURMAN J.A., SHEK J.W.: *Ann. Neurol.*, **8**, 479, 1980.
83. KIM Y.S., WISNIEWSKI H.M.: *Brain. Res.*, **377**, 286, 1986.
84. DU VAL G., GRUBB B.R., BENTLEY P.J.: *J. Toxicol. Environm. Health.*, **19**, 97, 1986.
85. SCHNEIDER D., PETERS H.-J., KOEHLER H., VORBERG B.: *Trace Elem. Med.*, **2**, 150, 1985.
86. BANKS W.A., KASTIN A.J., FASOLD M.B.: *J. Pharmacol. Exp. Ther.*, **244**, 579, 1988.
87. AUDUS K.L., SHINOGLE J.A., GUILLOT F.L., HOLTHAUS S.R.: *Int. J. Pharmacol.*, **45**, 249, 1988.
88. GLICK J.L.: *Med. Hypotheses*, **31**, 211, 1990.
89. PULLEN R.G.L., CANDY J.M., MORRIS C.M., TAYLOR G., KEITH A.B., EDWARDSON J.A.: *J. Neurochem.*, **55**, 251, 1990.
90. MARTIN R.B.: in *Metal Ions in Biological Systems*. SIEGEL H., SIGEL A., (Eds.), Marcel Dekker Inc., New York, vol. 24, 1988, pg. 1.
91. MARTIN R.B.: *J. Inorg. Biochem.*, **28**, 181, 1986.
92. ÖHMAN L.-O., SJÖBERG S.: *J. Chem. Soc. Dalton*, 2513, 1983.
93. VENTURINI M., BERTHON G., MAY P.M.: *Recueil des Travaux Chim. Pays-Bas*, **106**, 406, 1987.
94. MOLITORIS B.A., ALFREY A.C., MILLER N.L., et al.: *Kidney Int.*, **31**, 986, 1987.
95. AKESON M.A., MUNNS D.N.: *Biochim. Biophys. Acta*, **984**, 200, 1989.
96. PROVAN S.D., YOKEL R.A.: *J. Pharm. Exp. Ther.*, **245**, 928, 1988.
97. MOLITORIS B.A., FROMENT D.H., MCKENZIE T.A., HUFFER W.H., ALFREY A.C.: *Kidney Int.*, **36**, 949, 1989.
98. PEKELHARING J.M.: in *Lectin-Enzyme Binding Assays*. DRUKKERIJ J.H., PASMAN B.V.'s-Gravenhage, Rotterdam, 1989.
99. FAVARATO M., MIZZEN C., KRISHNAN B., KRUCK T.P.A., ZATTA P., MCLACHLAN D.R.: in *Alzheimer's Disease: Basic Mechanisms, Diagnosis and Therapeutic Strategies*. IQBAL K., MCLACHLAN D.R., WINBLAD B., WISNIEWSKI H.M. (Eds.). J. Wiley & Sons Ltd, Chichester, 1991, pg. 399.
100. WILLIAMS S.K., DEVENNY J.J., BITENSKY M.W.: *Proc. Natl. Acad. Sci-USA*, **78**, 2393, 1981.
101. PARDRIDGE M.W.: *Endocrine Rev.*, **7**, 314, 1986.
102. WILLIAMS S.K.: *New York Acad. Sci.*, **416**, 457, 1983.
103. KUMAGAI A.K., EISENBERG J., PARDRIDGE W.M.: *J. Biol. Chem.*, **262**, 15214, 1987.
104. HUEBERS H.A., FINCH C.A.: *Physiol. Rev.*, **6**, 520, 1987.
105. LIN H.L., CONNOR J.R.: *Dev. Brain Res.*, **49**, 281, 1989.
106. CHOSTEEN N.D.: *Coord. Chem. Rev.*, **22**, 1, 1977.
107. COCHRAN M., COATES J., NEOH S.: *FEBS-Letters*, **176**, 129, 1987.
108. MCGREGOR S.J., NAVES M.L., ORIA R., VASS J.K., BROCK J.H.: *Biochem. J.*, **272**, 377, 1990.
109. MILL J.M., RUFF M.R., WEBER R.J., PERT C.B.: *Proc. Natl. Acad. Sci-USA*, **42**, 4553, 1956.
110. CONNOR J.R., FINE R.E.: *Brain Res.*, **368**, 319, 1986.
111. JEFFRIES W.A., BRANDON M.R., HUNT S.V., WILLIAMS A.F., GATTER K.C., MASON D.Y.: *Nature*, **312**, 162, 1984.
112. FISHMAN R.E., RUBIN J.B., HANDROHAM J.V., CONNOR J.R., FINE R.E.: *J. Neurosc. Res.*, **18**, 299, 1987.
113. VAN DER VOET G.B., DE WOLFF F.A.: *Human Toxicol.*, **4**, 643, 1984.
114. TRAPP G.S.: *Life Sciences*, **33**, 311, 1983.
115. KING S.W., WILLS M.R., SAVORY J.: *Res. Comm. Chem. Pathol. Pharmacol.*, **26**, 161, 1979.
116. TODA W., LUX J., VAN LOON J.C.: *Anal. Letters*, **13(B13)**, 1105, 1980.
117. PIERIDES A.M., EDWARDS W.G., CULLUM V.X., MC CALL J.T., ELLIS H.A.: *Kidney Int.*, **18**, 115, 1980.
118. KING S.W., SAVORY J., WILLS M.R.: *Ann. Clin. Lab. Sc.*, **12**, 143, 1982.
119. FARRAR G., ALTMAN P., WELCH S., WYCHRIJ O., GHOSE B., LEJEUNE J., CORBETT J., PRASHER V., BLAIR J.: *Lancet*, **335**, 747, 1990.
120. MONTREUIL J., SPIK G.: in *Proteins and Iron Storage and Transport in Biochemistry and in Medicine*. CRICHTON R.R. (Ed.), North Holland, Amsterdam, 1975, pg. 27.

121. SPIK G., BAYARD B., FOURNET B., STRECKER G., BOUQUELET S., MONTREUIL J.: *FEBS-Letters*, **50**, 296, 1975.
122. SPIK G.: in *The Biochemistry and Physiology of Iron*. SALTMAN P., HEGENAUER J. (Eds.) Elsevier North Holland Inc., 1982, pg. 49.
123. HUNT R.C., RIEGLER R. DAVIS A.A.: *J. Biol. Chem.*, **264**, 9643, 1989.
124. VORBRODT A.W.: in *Ultrastructural Cytochemistry of Blood-Brain Barrier Endothelia*. Progress in Histochemistry and Cytochemistry vol. 18, n. 3. Gustav Fischer Verlag, Stuttgart, New York 1988.
125. CORAIN B., NICOLINI M., ZATTA P.: *Coord. Chem. Rev.* **000**, 000, 1991.
126. CORAIN B., TAPPARO A., SHEIK-OSMAN A.A., BOMBI G.G., ZATTA P., FAVARATO M.: *Coord. Chem. Rev.*, **000**, 000, 1991.
127. BOMBI G.G., CORAIN B., GIORDANO R., PAGANI D., SESTI A.G., ZATTA P.: in *Topics in Aging Research in Europe*, vol. 8 VEZZADINI P., FACCHINI A., LABO G., (Eds.) EURAGE, Rijswijk, 1986, pg. 253.
128. FINNEGAN M.M., RETTIG S.J., ORVIG C.: *J. Chem. Soc.*, **108**, 5033, 1986.
129. LEVIN V.A.: *J. Med. Chem.*, **23**, 682, 1980.
130. TAPPARO A., PERAZZOLO M.: *Int. Environm. Anal. Chem.*, **36**, 13, 1989.
131. HEM J.D.: *Kidney Int.*, **29** (Suppl. 18) S-3, 1986.
132. FONTANA L., PERAZZOLO M., STELLA M.P., TAPPARO A., CORAIN B., FAVARATO M., ZATTA P.: *Biol. Trace Elem. Res.*, **000**, 000, 1991.
133. BOMBI G.G., CORAIN B., SHEIK-OSMAN A.A., VALLE G.C.: *Inorg. Chim. Acta*, **171**, 79, 1990.
134. CORAIN B., LONGATO B., SHEIK-OSMAN A.A., BOMBI G.G., MACCA C.: *J. Chem. Soc. Dalton*. 000, 1991.
135. FAVARATO M., ZATTA P., NICOLINI M.: *II Int. Springfiled Symp. on Advances in Alzheimer Therapy*. (Abstract), p. 16, 1991.
136. ZATTA P., PERAZZOLO M., BOMBI G.G., CORAIN B., NICOLINI M.: in *Progress in Clinical and Biological Research*, Iqbal K., Wisniewski H.M., Winblsd B. (Eds.), Liss A. Inc., New York, 1988, pg. 1087.
137. BERTHOLF R.L., HERMAN M.M., SAVORY J., CARPENTER R.M., STURGILL B.C., KATSETOS C.D., VANDENBERG S.T., WILLS M.R.: *Toxicol. Appl. Pharm.*, **98**, 58, 1989.
138. KATZETOS C.D., SAVORY J., HERMAN N.M., CARPENTERS A., FRANKFURTER A., HEWITT C.D., WILLS M.R.: *Neuropath. Appl. Neurobiol.*, **16**, 511, 1990.
139. LANGUI D., ANDERTON B.H., BRION J.-P., ULRICH J.: *Brin Res.*, **438**, 67, 1988.
140. LEVIN S.: *Environm. Health Persp.*, **000**, 000, 1991.
141. FAVARATO M., ZATTA P., PERAZZOLO M., FONTANA L., NICOLINI M.: *Brain Res.* **000**, 000, 1991.
142. GOLDSTEIN I.J., PORETZ R.D.: in *The Lectins: Properties, Funtions and Applications in Biology and Medicine*. LIENER I.E., SHARON N., GOLDSTEIN I.J., (Eds.) Academic Press, New York, 1986, pg. 33.
143. VORBRODT A.W.: *Acta Neuropathol.*, **70**, 103, 1986.

COMMENTARY

About the Association of Aluminum and Alzheimer Disease - A Commentary

H.M. Wisniewski

Aluminum (Al) injected into the brain of experimental animals produces neurofibrillary changes, a term used by neuropathologists to describe conditions of excessive accumulations of tangles of fibrils in the nerve cells. There is also evidence that animals treated with Al have difficulties learning new tasks.[1,2]

Because neurofibrillary changes are seen in patients with certain types of dementias (presenile and senile dementia of the Alzheimer type - SDAT - and in Parkinsonism dementia amyotrophic lateral sclerosis complex - ALS-PD)[3] it has been suggested that Al could be the cause of these diseases.[4] In 1972, Alfrey and associates[5] described a new neurologic syndrome that appears in patients receiving chronic dialysis because of kidney failure. This syndrome, called *dementia dialytica*, is characterized by speech abnormality with subsequent severe memory impairment associated with an abnormal electroencephalogram with spike (seizure) activity. Further studies showed that in patients with dementia dialytica Al levels in the brain are very high and that this atom is probably responsible for this disease. Studies by Crapper et al.[6] demonstrated increased amounts of Al in neocortical samples of tissue from patients with Alzheimer disease (AD). Perl and associates[7] and Edwardson et al.[8] reported the presence of Al in the neurofibrillary tangles (NFT) and cores of amyloid stars. These findings taken together reinforced the belief that Al may be of importance in the etiology and/or pathogenesis of AD.

Our extensive studies on the pathology of SDAT and aluminum-induced encephalomyelopathy show that Al is not the cause of SDAT.[9] In our opinion, the different types of pathological and biochemical changes found in aluminum-treated animals and in humans with SDAT speak against the role of Al as an etiological agent in this disease.

The argument that differences in both the morphology and the biochemistry of neurofibrillary changes in AD and Al encephalopathy can be the result of species differences calls into question Foncin and El Hachimi's report.[10] In 1986, these authors reported a case of accidental metallic aluminum implantation in a human brain. The patient, asymptomatic for 15 years, afterward developed seizures and five years later died in status epilepticus. However, no neurofibrillary tangles or beta amyloid plaques were found in his brain; also, persons taking antacids containing large amounts of aluminum hydroxide showed no evidence of increased risk of developing AD.[11,12] Due mostly to methodological problems reports on the presence of Al in the NFT and amyloid cores are inconclusive[13-16] The suggestion that aluminum and silicon is the backbone on which the beta protein precipitates and forms amyloid fibrils[13,17] can be questioned by the same authors' data showing that aluminosilicate deposits are not a feature of amyloid angiopathy in AD, in spite of the fact that the vascular and core amyloid is made of the same beta-protein fibrils. Recent studies on the formation of beta-amyloid fibril deposits in the brains of transgenic mice[18,19] clearly showed that Al is not the cause of beta-amyloid fibril formation. Papers reporting the relationship between AD and aluminum in drinking water[20,21] also run into many methodological problems.[22]

In summary, there is no evidence that Al is responsible for AD type of neuropathology. However, under certain conditions when aluminum enters the brain, cognition can be affected and Al can cause dementia. In other words, Al is a dementing atom. Therefore, in the Summary Report of the Workshop on Aluminum and Health, held in Oslo in 1988,[23] the toxicology work group made the following recommendations to protect public health against the toxic effects of aluminum:

- Preterm infants are particularly vulnerable to the effects of Al because of their reduced renal function. Therefore the Al content of all intravenous solutions used in their management should be monitored, controlled, and labelled.
- The use of certain elemental additive solutions, used in total parenteral nutrition and in other circumstances, that are high in Al should be eliminated or reduced.
- The Al content of intravenous fluids used in children and adults with renal failure or undergoing dialysis should be less than 10 μg L^{-1} feeding fluids, wherever possible.
- Regular monitoring of plasma Al levels in uremic patients receiving Al-containing phosphate binders and/or dialysate is desirable. Al levels exceeding 80 μg L^{-1} should prompt physicians to consider the relative risks of phosphate accumulation and possible Al toxicity.
- Based on current data, serum Al levels should not exceed 80 μg L^{-1}. However, additional research is needed.
- Citrate-containing compounds appear to increase the bioavailability of oral Al, therefore, particular caution should be taken to avoid combining these compounds with Al-containing phosphate binders.

Acknowledgements. Supported in part by funds from the New York State Office of Mental Retardation and Developmental Disabilities and the fund for the Center for Trace Metal Studies and Environmental Neurotoxicology.

References

1. Wisniewski H.M., Narkiewicz O., Wisniewska K.: Topography and dynamic of neurofibriller degeneration in aluminum encephalopathy. *Acta Neuropathol.*, **9**, 127-133, 1967.
2. Rabe A., Lee M.H., Shek J., Wisniewski H.M.: Learning deficit in immature rabbits with aluminum-induced neurofibrillary changes. *Exp. Neurol.*, **76**, 441-446, 1982.
3. Hirano A., Zimmerman H.M.: Alzheimer's neurofibrillary changes: a topographic study. *Arch. Neurol.*, **7**, 227-242, 1962.
4. Crapper D.R., Krishnan S.S., Quittkat S.: Aluminum neurofibrillary degeneration and Alzheimer's disease. *Brain*, **99**, 67-80, 1976.
5. Alfrey A.C., Mishell M., Burks J., Contiguglia S.R., Rudolph H., Lewin E., Holmes J.H.: Syndrome of dyspraxia and multifocal seizure associated with chronic hemodialysis. *Trans. Am. Soc. Artif. Intern. Organs*, **18**, 257-261, 1972.
6. Crapper D.R., Krishnan S.S., Dalton A.J.: Brain aluminum distribution in Alzheimer's disease and experimental neurofibrillary degeneration. *Science*, **180**, 511-513, 1973.
7. Perl D.P., Good P.F.: Microprobe studies of aluminum accumulation in association with human central nervous system disease. *J. Environ. Geochem. & Health*, **12**, 97-101, 1990.
8. Edwardson J.A., Candy J.M.: Aluminum and the pathogenesis of senile plaques: studies in Alzheimer's disease and chronic renal failure. *J. Environ. Geochem. & Healt*, **12**, 94-96, 1990.
9. Wisniewski H.M., Moretz R.C., Sturman J.A., Wen G.Y., Shek J.W.: Aluminum neurotoxicity in mammals. *J. Environ. Geochem. & Health.*, **12**, 115-120, 1990.
10. Foncin J.F., El Hachimi K.H.: Neurofibrillary degeneration in Alzheimer's disease: a discussion with a contribution to aluminum pathology. In: *Senile Dementias: Early Detection*, Bes A., Cahn J., Cahn R., Hoyer S., Marc-Vergnes J.P., Wisniewski H.M. (Eds), John Libbey Eurotext, London-Paris, pp. 191-200, 1986.
11. Graves A.B., White E., Koepsell T.D., Reifler B.V., Van Belle G., Larson E.B.: The association between aluminum-containing products and Alzheimer's disease. *J. Clin. Epidemiology*, **43**, 35-44, 1990.
12. Broe G.A., Henderson A.S., Creasey H., McCusker E., Korten A.E., Jorm A.F., Longley W., Anthony J.C.: A case-control study of Alzheimer's disease in Australia. *Neurology*, **40**, 1698-1707, 1990.
13. Candy J.M., Oakley A.E., Klinowski J., Carpenter T.A., Perry R.H., Atack E.J.R., Perry E.K., Blessed G., Fairbairn A., Edwardson J.A.: Aluminosilicates and senile plaque formation in Alzheimer's disease. *Lancet*, **1**, 354-357, 1986.
14. Moretz R.C., Iqbal K., Wisniewski H.M.: Microanalysis of Alzheimer disease NFT and plaques. *J. Environ. Geochem & Health*, **12**, 15-16, 1990.
15. Perl D.P., Good P.F.: Microprobe studies of aluminum accumulation in association with human central nervous system disease. *J. Environ. Geochem. & Health*, **12**, 97-101, 1990.

16. MARKESBERY W.R., EHMANN W.D., HOSSAIN T.I.M., ALAUDDIN M., GOODIN D.T.: Instrumental neuron activation analysis of brain aluminum in Alzheimer disease and aging. *Annals Neurol.*, **10**, 511-516, 1981.
17. CANDY J.M., EDWARDSON J.A., KLINOWSKI J., OAKLEY A.E., PERRY E.K., PERRY R.H.: Co-localization of aluminum and silicon in senile plaques: implications for the neurochemical pathology of Alzheimer's disease. In: *Senile Dementia of the Alzheimer Type: Early Diagnosis, Neuropathology and Animal Models*, Traber J., Gispin W.H. (Eds), Springer-Verlag, Berlin, pp. 183-197, 1985.
18. QUON D., WANG Y., CATALANO R., SCARDINA J.M., MURAKAMI K., CORDELL B.: Formation of β-amyloid protein deposits in brains of transgenic mice. *Nature*, **352**, 239-241, 1991.
19. WIRAK D.O., BAYNEY R., RAMABHADRAN T.V., FRACASSO R.P., HART J.T., HAUER P.E., HSIAU P., PEKAR S.K., SCANGOS G.A., TRAPP B.D., UNTERBECK A.J.: Deposits of amyloid β protein in the central nervous system of transgenic mice. *Science*, **253**, 323-325, 1991.
20. FLATEN T.P.: Geographical associations between aluminum in drinking water and death rates with dementia (including Alzheimer's disease), Parkinson's disease and amyotrophic lateral sclerosis in Norway. *J. Environ. Geochem. & Health*, **12**, 152-167, 1990.
21. MARTYN C.N.: Aluminum and Alzheimer's disease: An epidemiological approach. *J. Environ. Geochem. & Health*, **12**, 169-171, 1990.
22. SCHUPF N., SILVERMAN W., ZIGMAN W.B., MORETZ R.C., WISNIEWSKI H.M.: Aluminum and Alzheimer's disease. *Lancet*, **1**, 267, 1989.
23. CONNERY J.: Summary report workshop on aluminum and health, Oslo, May 2-5, 1988. *J. Environ. Geochem. & Health*, **12**, 179-187, 1988.

PRINTED IN ITALY
BY BERTONCELLO ARTIGRAFICHE,
CITTADELLA (PADOVA)
SEPTEMBER 1991